T0220259

Ruby Data Processing

Using Map, Reduce, and Select

Jay Godse

Apress®

Ruby Data Processing: Using Map, Reduce, and Select

Jay Godse
Kanata, Ontario, Canada

ISBN-13 (pbk): 978-1-4842-3473-0   ISBN-13 (electronic): 978-1-4842-3474-7
https://doi.org/10.1007/978-1-4842-3474-7

Library of Congress Control Number: 2018934400

Managing Director, Apress Media LLC: Welmoed Spahr
Acquisitions Editor: Steve Anglin
Development Editor: Matthew Moodie
Coordinating Editor: Mark Powers

Cover designed by eStudioCalamar

Cover image designed by Freepik (www.freepik.com)

Distributed to the book trade worldwide by Springer Science+Business Media New York, 233 Spring Street, 6th Floor, New York, NY 10013. Phone 1-800-SPRINGER, fax (201) 348-4505, email orders-ny@springer-sbm.com, or visit www.springeronline.com. Apress Media, LLC is a California LLC and the sole member (owner) is Springer Science + Business Media Finance Inc (SSBM Finance Inc). SSBM Finance Inc is a **Delaware** corporation.

For information on translations, please email editorial@apress.com; for reprint, paperback, or audio rights, please email bookpermissions@springernature.com.

Apress titles may be purchased in bulk for academic, corporate, or promotional use. eBook versions and licenses are also available for most titles. For more information, reference our Print and eBook Bulk Sales web page at http://www.apress.com/bulk-sales.

Any source code or other supplementary material referenced by the author in this book is available to readers on GitHub via the book's product page, located at www.apress.com/9781484234730. For more detailed information, please visit http://www.apress.com/source-code.

Printed on acid-free paper

Table of Contents

About the Author

Jay Godse is an active software and web applications developer with expertise in Ruby, Rails, various databases, and Ansible. He also is active on Stack Overflow as an active contributor. He graduated with an engineering degree and then went to work as a digital circuit designer. After a year of that, he switched to software development, and he has been there ever since in some form. His early work was mostly real-time telecommunication device control and provisioning using languages such as C and Protel. He then transitioned into designing distributed computing systems using languages such as C++. After that, he moved into web applications. Along the way, he did stints as a software development manager and a software architect. But for the last nine years, he has written web applications in Ruby and DevOps applications in Ansible and Powershell.

About the Technical Reviewer

Massimo Nardone has more than 24 years of experience in security, web/mobile development, cloud, and IT architecture. His true IT passions are security and Android.

He has been programming and teaching how to program with Android, Perl, PHP, Java, VB, Python, C/C++, and MySQL for more than 20 years.

He holds a Master of Science in Computing Science from the University of Salerno, Italy.

He has worked as a project manager, software engineer, research engineer, chief security architect, information security manager, PCI/SCADA auditor, and senior lead IT security/cloud/SCADA architect for many years.

Technical skills include security, Android, cloud, Java, MySQL, Drupal, Cobol, Perl, web and mobile development, MongoDB, D3, Joomla, Couchbase, C/C++, WebGL, Python, Pro Rails, Django CMS, Jekyll, Scratch, and more.

He worked as visiting lecturer and supervisor for exercises at the Networking Laboratory of the Helsinki University of Technology (Aalto University). He holds four international patents (PKI, SIP, SAML, and Proxy areas).

He currently works as chief information security officer (CISO) for Cargotec Oyj and is member of ISACA Finland Chapter Board.

Massimo has reviewed more than 45 IT books for different publishers, and in addition to reviewing this book he is also the coauthor of *Pro Android Games* (Apress, 2015).

Acknowledgments

I would like to acknowledge a few people who helped make me a better programmer and a better writer. Thanks go to Kevin Szabo and Ronnie Taylor, both of whom helped me be a better programmer, and to Christina Hardy, who helped me to become a better writer. And thanks also go to Mark Powers and the publication team at Apress who helped produce this book.

Introduction

I wrote mostly reactive software for many years, but there was always a user interface or reporting component that had string manipulation, data synthesis, or data formatting. Since I was not trained as a computer scientist, I did not learn of higher-order functions such as map(), reduce(), and select() **that were** found in languages like Lisp or Smalltalk. Also, Ruby, Python, C++, and JavaScript were not around when I was in school. As a result, I struggled with cumbersome, error-prone imperative code for some tasks.

A few years after learning Ruby, but while still using the imperative programming style for data processing, I discovered the Ruby Enumerable library and started using its higher-order functions, such as map(), reduce(), and select(). What happened to my data-processing code?

- Code became more straightforward because data-processing tasks could be expressed with a cascading pipeline of **map, reduce**, and **select** stages. Each stage in the pipeline was simple to develop and debug.

- Code volume was reduced by half.

- Code was naturally more robust, and debugging took much less time.

- By thinking of solutions in terms of **map, reduce**, and **select**, I was able to envision and solve more complex problems.

However, I couldn't find any but the most trivial examples of how to use these functions to solve data-processing problems. It took me a lot of time-consuming trial and error with **map, reduce,** and **select** to solve these kinds of problems.

I decided to write this book to

- codify my learning and help you understand it more quickly and effectively than I did; and

- show you my problem-solving approaches to help you solve your data-processing problems.

I'll also admit that I was forced to learn many of the programming nuances of these functions as I wrote the examples.

Who This Book Can Help

This book can help those who

- want to become more skilled in writing data-processing code by becoming more fluent in using **map, reduce** and **select**;

- are willing to type in all the code in the examples and run the code using the **irb** Ruby command line and a text editor;

- are willing to carefully read and try the examples in this book and to reflect and ponder the outputs, and even tweak the programs independently to try to understand what is happening;

- are open to learning a different way to approach data-processing problems; and

- are willing to try failed approaches to learn the nuances of these functions.

Who This Book Might Not Help

This book might not help you if you want

- recipes for standard problems via a cookbook; or

- a reference for data-processing solutions.

Prerequisites

You should be familiar with some Ruby programming or at least some programming in another language. If you don't know Ruby, search out the book *Learn Ruby the Hard Way*[1] and go through it as prescribed by its author, Zed Shaw.

If you are fluent with Python, you might benefit from this book if you go slowly and look up Ruby information online as you move through this book.

An internet connection helps if you want to search for online help using Google, Bing, or another search engine.

You should have a computer with Ruby 2.2.x installed. Windows works, as do Linux and Mac OSX.

You should have a good syntax-highlighting text editor. I recommend Notepad++ on Windows or gedit on Linux or Macintosh, both of which are free. Sublime works on Windows, Linux, or Macintosh and is slightly better, but costs about $70 at the time of this writing.

---

[1]http://learnrubythehardway.org/book/

CHAPTER 1

Basic Ruby

This section will acquaint or refresh you with basic ways to use the Ruby command line, as well as some relevant Ruby coding.

If you are comfortable programming in Ruby and understand the Ruby Enumerable Library reasonably well, you can skip this section.

The Command Line

After you install Ruby, you can fire up the interactive command line in Windows, Linux, or Mac OSX. I'm using Windows.

```
C:\>  irb
irb(main):001:0>
```

For brevity, I won't write the full irb prompt every time.

You can execute Ruby statements line by line. The value returned by each expression is preceded by ⇒.

```
irb> a = 4
=> 4
irb> a
=> 4
irb> a + 7
=> 11
```

© Jay Godse 2018
J. Godse, *Ruby Data Processing*, https://doi.org/10.1007/978-1-4842-3474-7_1

You can choose to put multiple statements on one line, separated by a semicolon.

```
irb> list = ["one", "two","three"];1
=>1
irb> list
=>["one", "two", "three"]
```

You can also have a statement that spans multiple lines.

```
irb> [2,3,4,5].each do |n|
irb>     if n%2==0
irb>         puts "even"
irb>     else
irb>         puts "odd"
irb>     end
irb> end
=> "even"
=> "odd"
=> "even"
=> "odd"
```

You can put the code into a file called sample.rb, located in the same directory or folder from which you ran irb.

```
1   [2,3,4,5].each do |n|
2       if n%2 == 0
3           puts "even"
4       else
5           puts "odd"
6       end
7   end
```

Then, with `irb`:

```
irb> load "sample.rb"
irb> end
=> "even"
=> "odd"
=> "even"
=> "odd"
irb>
```

You could also copy the code block from your text editor and paste it directly into your command line and get the same result, as long as you use only spaces for indentation. (Using tabs for indentation will work fine if you load the file from the command prompt, but if you paste tabs directly into `irb`, it will generate errors).

You can either type the code samples from this book into `irb` directly or type them into a text editor and then load the file as just shown.

Object Scope

When you are in a Ruby program, the general method of executing a method f on an object `obj` is as follows:

```
obj.f
```

That is true whether you are in a Ruby program or the command line. When code is in the main program file, or in the `irb` command-line interpreter, there is an implied context for many functions, such as `puts`. That is why you can do this:

```
puts "Hello world"
```

You don't need to specify a class to qualify `puts`, because `puts` is a method of the underlying context object.

String

Strings are basic constructs in all languages. Let's look at a few basic operations used in this book. Try them out on the irb command line for yourself.

length or size

> This yields the size of the string.

downcase

> This converts all letters to lowercase.

upcase

> This converts all letters to uppercase.

capitalize

> This capitalizes the first letter of a string.

split()

> This searches a substring for the argument string and splits the string into a array comprising substrings on both sides of the split argument, while the substring of the split argument is discarded. If there are no matches, an array is returned with the whole string:

```
irb> base_string = "abc def  ghi"
irb> base_string.split(" ")
=> ["abc","def","ghi"]
irb> base_string.split("  ")
=> ["abc def","ghi"]
irb> base_string.split(" d")
=> ["abc","ef ghi"]
irb> base_string.split("efghi")
=> ["abc def ghi"]
```

join()

This joins each element of an array of strings with
the string in the join argument as the separator.

```
irb> stringlist = ["This", "is", "a", "sentence"]
irb> stringlist.join(" ")
=> "This is a sentence"
irb> stringlist.join(",")
=> "This,is,a,sentence"
irb> stringlist.join(" A SPACE ")
=> "This A SPACE is A SPACE a A SPACE sentence"
```

string interpolation

This lets you define a string template with
parameters. The string #{} gives you a place to put a
variable.

```
irb> first = "Jack"; last = "Black"
irb> full_name = "#{first} #{last}"
=> "Jack Black"
```

This is especially useful for iterating over a list:

```
irb> smith_brothers = ["Terry", "Jerry", "Harry"]
irb> smith_brothers.each{|brother| puts "#{brother} Smith"}
Terry Smith
Jerry Smith
Harry Smith
=> ["Terry", "Jerry", "Harry"]
```

C/C++ programmers will recognize string
interpolation as being similar to the sprintf
function in the C standard library.

Array

Arrays in Ruby are like arrays in other languages. They are a collection of things indexed by a whole number (e.g., 0,1,2,...). In Ruby, an array can contain any Ruby object at an index. Array indices in Ruby start at 0.

Arrays implement the Ruby Enumerable interface, so they will have key methods such as each, map, reduce, select, and others.

Special Methods

compact()

> This method on an array gets rid of nil members. For example:
>
> ```
> irb> [1,nil,2,nil,nil,3,nil,[]].compact
> => [1,2,3,[]]
> ```

flatten()

> This method gets rid of inner arrays to an arbitrary depth. It is often useful when dealing with nested map and reduce statements. For example:
>
> ```
> irb> [[1,[2,3]],4,[5,6]].flatten
> => [1,2,3,4,5,6]
> ```

push()

> This method pushes an element onto the end of an array and returns a new array. For example:
>
> ```
> irb> [1,2,3,4,5].push(666)
> => [1,2,3,4,5,666]
> ```

pop

This method removes the last element of an array
and returns that element and returns a new array.
For example:

```
irb> the_array = [1,2,3,4,5]
irb> tail = the_array.pop
irb> tail
=> 5
irb> the_array
=> [1, 2, 3, 4]
```

unshift()

This method pushes an element onto the beginning
of an array and returns a new array. For example:

```
irb> the_array = [1,2,3,4,5]
irb> the_array.unshift(666)
=> [666, 1, 2, 3, 4, 5]
```

shift

This method removes the first element of an array
and returns that element and modifies the old array.
For example:

```
irb> the_array = [1,2,3,4,5]
irb> head = the_array.shift
irb> head
=> 1
irb> the_array
=> [2, 3, 4, 5]
```

Hash

Another term for hash is an "associative array," or even a "dictionary,"
Hashes are indexed by a key object, and there is a value (another object)
for each key object.

```
irb> newhash = Hash.new
=> {}
irb> newhash2 = {}
=> {}
```

These are two ways of creating a new hash. The key of a hash can be
any object or symbol.

```
irb> newhash = {key1: "some_value", key2: 2}
=> {:key1=> "some_value", :key2=>2}
irb> newhash2 = {}
irb> newhash2[:key1] = "key1_symbol"
irb> newhash2["key1"] = "key3_string"
irb> newhash2[3] = 3
irb> newhash2[:key4] = [4,"44"]
irb> newhash2[[3,"33"]] = "wow"
irb> newhash2
=> {:key1=>"key1_symbol", "key2"=>"key2_string", 2=>1,
:key4=>[4,"44"], [3,"33"]=\
"wow" }
```

The key can be any object, including symbols, strings, numbers,
hashes, arrays, or anything. When iterating through a hash using each or
map, it emits a single array with the key as the first element and the value
as the second element, or it can emit two elements, which are the key and
the value. (Note that inspect is how you can inspect the content of any
object).

```
irb> newhash = {a: 1, b: 2, c: 3}
irb> newhash.each{|e| puts e.inspect}
[:a, 1]
[:b, 2]
[:c, 3]
=> {:a => 1, :b => 2, :c => 3}

irb> newhash.map{|key, value| "The key is #{key} and the value
is #{value}"}

=> ["The key is a and the value is 1", "The key is b and the
value is 2", "The key is c and the value is 3"]
```

Block-passing Syntax

Ruby is one of many languages that allow lexical closures, otherwise
known (kind of) as anonymous functions or blocks. These functions are
dynamically created. The function can take the form of a defined function.
For example:

```
irb> def plus_two(a,b)
irb>    a + b + 2
irb> end
irb> plus_two(3,4)
=> 9
```

Or, you could define it as follows:

```
irb> plus_two = ->(a,b) do
irb>    a + b + 2
irb> end
```

Or, you could define it this way:

```
irb> plus_three = ->(a,b){a + b + 3}
irb> plus_three.call(5,7)
=> 15
```

Now, anonymous functions or blocks can be passed to other functions as parameters. A function that takes a function (defined or anonymous) as a parameter will optionally execute the function in its program flow. Suppose there is a function called printit() defined as follows:

```
irb> printit = ->(a) do
irb>     puts a.inspect
irb> end
```

Now, suppose you want to print each element of a range. You will call each on the array, which will run the block passed to each for each element of the array.

```
irb> (1..5).each(&printit)
1
2
3
4
5
=> (1..5)
```

You could also pass an actual block of code to run. The block is surrounded by **do** and **end** or { and }. Objects between the vertical bars are optionally passed to the block (depending on the definition of the block), and they are used by the block to execute. The return value of a block is the value of the last expression in the block.

```
irb> (1..5).each do |num|
irb>     puts num.inspect
irb> end
```

10

You will get the same output as earlier. Ditto for the following:

```
irb> (1..5).each{|num| puts num.inspect }
```

In Ruby, map, reduce, and select all take a code block as a parameter, and the code block is executed as defined by the function. In this book, I will use the do-end syntax most often, and sometimes the { } syntax.

Reading from Files

A common operation in data processing is reading from files and writing to files. Suppose you have names.txt, which looks like this:

```
1,John
2,Jack
3,Jim
4,Jared
5,John
```

To read it, you would do the following:

```
irb> names = File.open("names.txt").read
```

Now, on some computers, new lines are split by a carriage return ("\n"), and on other computers lines are split by a line feed and a carriage return ("\r\n").

If you want to split this file into rows, you would do the following:

```
irb> names.split("\n")
```

That could yield

```
=>["1,John\r", "2,Jack\r", "3,Jim\r", "4,Jared\r", "5,John\r"]
```

or

```
=>["1,John", "2,Jack", "3,Jim", "4,Jared", "5,John"]
```

To ensure consistency, use the chomp method, which removes whitespace characters if they exist. For example:

```
irb> "John\r".chomp
=> "John"
irb> "John".chomp
=> "John"
```

So, for our array, just to ensure that we don't pick up stray line feeds, we do the following:

```
irb> names.split("\n").map{|row| row.chomp}
=>["1,John", "2,Jack", "3,Jim", "4,Jared", "5,John"]
```

I assume that the examples in this book don't use "\r\n", but rather just "\n". If that doesn't work on your operating system, use the chomp() method as shown before working with the array.

CHAPTER 2

Function Overview and Simple Examples

The Ruby library includes the module Enumerable. This library contains `map()`, `reduce()`, `select()`, and other functions. This section will outline the syntax and meanings of the different parts of code that use these three functions.

If you can get through this section comfortably, both typing the code into `irb` and understanding the results, then you will be in a good position to deepen your understanding with the complex examples and reverse engineering that follow in the next chapters.

Map

This function of the Ruby Enumerable library is simple but profound. The `map()` method is applied to an array or a hash. The job of `map()` is to apply a function or block to each member of the array and return a new array.

So, when you see

```
def f(x)
    x*x
end
```

© Jay Godse 2018
J. Godse, *Ruby Data Processing*, https://doi.org/10.1007/978-1-4842-3474-7_2

```
output_array =
[1,2,3,4,5].map do |number|
    f(number)
end
```

you read, for each number in [1,2,3,4,5] apply f(x) to return the array [f(1), f(2), f(3), f(4), f(5)]. In this case, the answer is [1,4,9,16,25].

One could encode it in traditional imperative programming as follows:

```
output_array = []
for number in 1..5
    output_array.push( f(number) )
end
```

Using map, for example:

```
output_array =
[1,2,3,4,5].map do |number|
    number*number
end
```

Or, you could encode it as follows:

```
output_array =
[1,2,3,4,5].map{|element| element*element}
```

[1,2,3,4,5] is the array. The function (or block) to be applied to each element is {|element| element*element }.

With the { } syntax, or the do-end syntax, the object in the vertical bars (represented by element) is the element of the array currently being acted upon by the block, and the last object in any function or block is the return value, so element*element is returned. The net result is the following:

```
[1,4,9,16,25]
```

Suppose you wanted to provide data to a graphing program that plotted the square of the value.

```
[1,2,3,4,5].map{|element| [element, element*element]}
```

In this case, each "point" is a tuple (array) with an element and its square. This returns the following:

```
[[1,1],[2,4],[3,9],[4,16],[5,25]]
```

The difference is that the block returns an array each time with the number and its square, so the result is an array of arrays.

Since map returns an array, you can cascade map calls. For example, you could have a map block square each element, and then a second map block add 100. For example:

```
[1,2,3,4,5].map do |number|
    number*number
end.map do |square|
    square + 100
end
```

```
=> [101, 104, 109, 116, 125]
```

A number of the more complex operations can be solved by cascading with map calls.

Reduce

This function of the Ruby Enumerable library is more complex than map() and is quite powerful. The method is applied to a collection (for example, an array or a hash). The job of reduce() is to apply a function cumulatively to each member of the collection and then return an object (which could be an array, a single value, a hash, or anything).

For example:

```
[1,2,3,4,5].reduce(0, :+)
```

This is like saying for each member of the list, add (:+) the member to memo and save the memo for the next element. The initial value of memo is 0 (the first parameter). Another way to code it is as follows:

```
memo = 0
[1,2,3,4,5].each{|element| memo = memo.+(element)}
```

Or, you can pass reduce a block of code to run. The block comes after the closing parenthesis and is surrounded by a do-end construct, or a {} construct. The emitted variables are always the memo first and the element second. (In this case, the initial value of memo is 33).

```
[1,2,3,4,5].reduce(33) do |memo, element|
    memo.+(element)
end
memo
```

```
=> 48
```

(You don't have to call them "memo" or "element," but that is how they behave). After each iteration, the memo takes the value of the last object in the block. For example:

```
[1,2,3,4,5].reduce(33) do |memo, element|
    memo.+(element)
    77.7
end
```

```
=> 77.7
```

This returned 77.7, because that was the last object in the block.

Another thing to remember is that the memo returned by the block has to be of the same class as the memo emitted by the reduce function. For example, when you are using reduce() to build a Hash object, you must return the whole Hash object:

```
[1,2,3,4,5].reduce({}) do |memo, element|
    memo[element] = element.to_s
end
```

```
IndexError: index 2 out of string
```

There is an error because the first run of the iteration returned memo[element], which is a String class, but the input expects it to be a Hash. To correct it, make sure that the memo of class Hash is explicitly the last object returned, as follows:

```
[1,2,3,4,5].reduce({}) do |memo, element|
    memo[element] = element.to_s
    memo
end
```

```
=> {1=>"1", 2=>"2", 3=>"3", 4=>"4", 5=>"5"}
```

This worked because memo is a Hash class and was returned every time.

You can think of the reduce operation as operating on each element and saving some kind of memo to carry forward to the operation on the next element. In the preceding example, where you calculate the sum of five numbers,

```
[1,2,3,4,5].reduce(0,:+)
```

or

```
[1,2,3,4,5].reduce(0) do |memo, element|
    memo + element
end
```

you are initializing the remembered quantity with a 0 and telling it to remember to add the current element to the previously reduced quantity and save it. The initial value of memo is an integer, and the last object returned by the reduce block is also an integer.

You could remember other things in the memo too. For example, you could save the last two numbers of a sequence. You would start with an initial value of [nil,nil] for the last two remembered things. On every iteration, you would push the element onto the memo (end of the array), and then you would shift off the first element of the array.

```
initial_memo = [nil,nil]
last_2 =
(1..10).reduce(initial_memo) do |memo, element|
    initial_memo.push( element )
    initial_memo.shift
    initial_memo
end

=> [9,10]
```

In general, if you have to remember something about the previously used elements when operating on the current element, use reduce(). If you don't need to remember anything, use map().

Sometimes reduce returns an array, in which case you can cascade it with other reduce calls or map calls. For example, if you want to add the sum of squares, you can use a map call to square each element, and then a reduce call to add them. For example:

```
[1,2,3,4,5].map do |number|
    number*number
end.reduce(0) do |sum, square|
    sum + square
end

=> 55
```

Or, perhaps you want to find the odd numbers of the last five in a sequence:

```
sequence = [1,2,3,455,5,6,4,3,45,66,77,54,23,4,55,6,7]
sequence.reduce([nil,nil,nil,nil,nil]) do |memo, number|
    memo.push(number)
    memo.shift
    memo
end.select do |one_of_last_5|
    one_of_last_5%2 == 1
end
=> [23, 55, 7]
```

If this seems complex and idiomatic, don't worry, because we'll explore these programming mechanics later in the book. It suffices to say that reduce can be used on both sides of a data-processing cascade. And please always remember that the input memo must be of the same class or structure as the output memo.

Simple Reduce Examples

I'll show you here how to implement some functions already available for arrays, but using reduce(), map(), and select().

uniq

This function finds a set of unique elements in an array. For example:

```
irb> [1,3,5,6,7,8,6,6,1,1].uniq
=> [1,3,5,6,7,8]
```

To implement uniq, we need a way to identify unique elements. First, realize that a hash Hash is keyed by unique values. So, as we run reduce, we just have to put each element as the key of a hash. If the element already exists, the value in the hash is overwritten. So, the initial value of the memo should be {}.

```
[1,3,5,6,7,8,6,6,1,1].reduce({}) do |memo, element|
    memo[element]=element
    memo
end
```

⇒ {1⇒1, 3⇒3, 5⇒5, 6⇒6, 7⇒7, 8⇒8}

Almost. Now, we just have to pick off the keys using map:

```
[1,3,5,6,7,8,6,6,1,1].reduce({}) do |memo, element|
    memo[element]=element
    memo
end.map do |key, value|
    key
end
```

=> [1, 3, 5, 6, 7, 8]

reverse

This function reverses the order of the elements in an array:

```
irb> [1,3,5,6,7,8,6,0].reverse
```

=> [0, 6, 8, 7, 6, 5, 3, 1]

One way to solve this is to keep pushing each element on a stack. A *stack* is just an array where you use unshift to push an element onto the front, and shift to pop elements off the front. The initial value of the reduce memo is a blank array. I put a print statement so you can see for yourself how the memo builds up.

```
[1,3,5,6,7,8,6,0].reduce([]) do |memo, element|
    memo.unshift element
    puts "After unshifting element: #{element}, the memo is:
    #{memo.inspect}"
    memo
end
```

The output is as follows:

```
After unshifting element: 1, the memo is: [1]
After unshifting element: 3, the memo is: [3, 1]
After unshifting element: 5, the memo is: [5, 3, 1]
After unshifting element: 6, the memo is: [6, 5, 3, 1]
After unshifting element: 7, the memo is: [7, 6, 5, 3, 1]
After unshifting element: 8, the memo is: [8, 7, 6, 5, 3, 1]
After unshifting element: 6, the memo is: [6, 8, 7, 6, 5, 3, 1]
After unshifting element: 0, the memo is: [0, 6, 8, 7, 6, 5, 3, 1]
=> [0, 6, 8, 7, 6, 5, 3, 1]
```

max

This function finds the largest element in an array:

```
irb> [1,3,5,6,7,8,6,0].max

=> 8
```

Reduce will work here because as you iterate through the array, you save the largest value in the memo. This example does not even need an explicit initial value, because the first element is the default value for the memo.

```
[4,3,5,6,7,8,6,0].reduce do |memo, element|
    if element > memo
        memo = element
    else
        memo = memo
    end
    puts "The largest element after checking #{element} is #{memo}"
    memo
end

The largest element after checking 3 is 4
The largest element after checking 5 is 5
The largest element after checking 6 is 6
The largest element after checking 7 is 7
The largest element after checking 8 is 8
The largest element after checking 6 is 8
The largest element after checking 0 is 8
=> 8
```

Of course, the terse way to encode this is as follows:

```
[4,3,5,6,7,8,6,0].reduce do |memo, element|
    (element > memo) ? element : memo
end

=> 8
```

Select

This function of the Ruby Enumerable library is very simple. The select()
method is applied to an array or a hash. The job of select() is to select
elements that return true for a predicate applied to the element. (A *predicate*
is an expression that returns true or false, and it runs in the block passed
to select).

The following trivial example has the predicate always return true,
and therefore all members of the set are selected. In this case, the set is
[1,2,3,4,5] and the predicate is true:

```
[1,2,3,4,5].select do |element|
    true
end
```

```
=> [1,2,3,4,5]
```

A slightly less trivial example selects even numbers (i.e., where the
predicate is **element % 2 == 0 **) from the original set of [1,2,3,4,5]:

```
[1,2,3,4,5].select do |element|
    element % 2 == 0
end
```

```
=> [2,4]
```

Selects can be cascaded. Suppose you want to find the even numbers
between 1 and 100 starting with a 7. First, you select the even numbers.
Then, you select the numbers starting with a 7:

```
irb>  (1..100).select{|number| number % 2 == 0}\
irb*> .select{|even_number| even_number.to_s[0] == "7"}
=> [70,72.74,76,78]
```

23

You could also reverse the cascade:

```
irb> (1..100).select{|number| number.to_s[0] == "7"}\
irb*> .select{|seven_number| seven_number % 2 == 0}
=> [70,72.74,76,78]
```

Since `select` returns an array, you can cascade `select` calls. For example, you could have a `select` block select even numbers and then square them using a `map` call. For example:

```
[1,2,3,4,5].select do |number|
    number%2==0
end.map do |even_number|
    even_number * even_number
end

=> [4, 16]
```

A number of the more complex operations can be solved by cascading with `select` calls.

CHAPTER 3

Complex Solutions

This section's examples will show you how to approach and solve more complex problems using map, reduce, and select. I will start off with a section on debugging and then follow it with complex solutions. Please go through the debugging section first because this is the key to understanding your problem and to developing a better understanding of map, reduce, and select.

Debugging Blocks for Map, Reduce, and Select

The best way to learn about your problem as you solve it is to insert print statements like puts to see what happens as you execute the loop.

Debugging Map Blocks

The thing to remember is that map operates on each element of the collection for each run of the block. When debugging, you want to see how the block operated on each element.

```ruby
[1,2,3,4,5].map do |number|
    result = number*number
    puts "The result of the function on element: #{element} is
    #{result}"
    result
end
```

It is important to save the result in a local variable so that you can return it as the last statement of the block. If you don't, the result of the puts statement will be returned (usually nil).

Debugging Reduce Blocks

Debugging reduce is a bit different because the memo is carried from element to element. Therefore, you want to print out the element as well as the memo.

```
final_answer =
[1,2,3,4,5].reduce(0) do |memo, number|
    memo = memo + number
    puts "After operating on number: #{number}, the memo is
    #{memo.inspect} "
    memo
end

After operating on number: 1, the memo is 1
After operating on number: 2, the memo is 3
After operating on number: 3, the memo is 6
After operating on number: 4, the memo is 10
After operating on number: 5, the memo is 15
=> 15
```

Again, make sure that you save the memo before printing it, then return it at the end of the block.

If you are using memo to build a different data structure as a part of the reduce block, inspecting the memo becomes extremely useful in converging on a good solution.

Debugging Select Blocks

Debugging select is easy because the only thing you need to know is whether the predicate was true for a particular element.

For example, a select filter that only keeps even numbers (number%2==0) could be debugged like this:

```
[1,2,3,4,5].select do |number|
    predicate = number%2==0
    puts "The predicate is #{predicate} for number #{number}"
    predicate
end
```

```
The predicate is false for number 1
The predicate is true for number 2
The predicate is false for number 3
The predicate is true for number 4
The predicate is false for number 5
=> [2, 4]
```

Here, you can see that only the numbers for which the predicate was true ended up in the answer ([2,4]).

FizzBuzz

This program is used as a programming question. Write a program to go through a list of integers 1..N. If the number is divisible by 2, print "fizz"; if it is divisible by 3, print "buzz"; if it is divisible by 2 and 3, print "fizzbuzz"; otherwise, print the number.

The input can be modeled as a range: (1..15).

The output should look like this:

```
1
fizz
buzz
fizz
5
fizzbuzz
7
fizz
buzz
fizz
11
fizzbuzz
13
fizz
buzz
```

The data structure for the output should be an array. We want to do something to each element of the array to get the output. Let's start with *fizz*, which requires that the number be divisible by 2 (n%2==0):

```
(1..15).map do |n|
    if n%2 == 0
        "fizz"
    else
        n.to_s
    end
end
```

That gives us the following:

```
=> ["1", "fizz", "3", "fizz", "5", "fizz", "7", "fizz", "9",
"fizz", "11", ... \ and so on
```

That is close, but we didn't address the *buzz*:

```ruby
(1..15).map do |n|
  if n%2 == 0
    "fizz"
  elsif n%3 ==0
    "buzz"
  else
    n.to_s
  end
end
```

That gives us the following:

```ruby
=> ["1", "fizz", "buzz", "fizz", "5", "fizz", "7", "fizz",
"buzz", "fizz", "11",\
"fizz", "13", "fizz", "buzz"]
```

That's close, but the *fizzbuzz* examples didn't happen at 6 and 12. Let's add in another clause. To be divisible by 2 and 3 implies that it must be divisible by 6:

```ruby
fizzbuzz_list =
(1..15).map do |n|
  if n%2 == 0
    "fizz"
  elsif n%3 ==0
    "buzz"
  elsif n%6 == 0
    "fizzbuzz"
  else
    n.to_s
  end
end
```

This yielded the same result as the previous attempt. A closer inspection of the code suggests that if the number is divisible by 6, it will always be even and will be caught by the n%2==0 clause. However, if we catch the n%6==0 clause first, what happens?

```ruby
fizzbuzz_list =
(1..15).map do |n|
    if n%6 == 0
        "fizzbuzz"
    elsif n%3 ==0
        "buzz"
    elsif n%2 == 0
        "fizz"
    else
        n.to_s
    end
end
```

That yields the following:

```
=> ["1", "fizz", "buzz", "fizz", "5", "fizzbuzz", "7", "fizz",
"buzz", "fizz", "\
11", "fizzbuzz", "13", "fizz", "buzz"]
```

To print it, just do the following:

```ruby
fizzbuzz_list.each{|e| puts e}
```

Sum of Odd Cubes

This is a typical math problem that can be solved using map, reduce, and select.

You need to sum the cubes of the odd numbers from 1 to 1000. For example, for the odd numbers under 10, it would be as follows:

```
1**3 + 3**3 + 5**3 + 7**3 + 9**3

=> 1225
```

1225 makes sense because it is 1 + 27 + 125 + 343 + 729.

How does one solve it? Start in stages. First, we have to figure out how to filter the odd numbers. An odd number is one where when you do a modulo-2 operation, it returns 1. For example:

```
num = 15
num%2

=> 1

num = 1000
num%2

=> 0
```

We could try selecting numbers from a range. For the purposes of this exercise, I'm going to restrict myself to numbers under 20.

```
(1..20).select do |number|
    number%2 == 1
end

=> [1, 3, 5, 7, 9, 11, 13, 15, 17, 19]
```

So far, so good. We have selected only odd numbers. The next step is to cube each of those numbers. In other words, we have to map each of those numbers to its cube.

```
(1..20).select do |number|
    number%2 == 1
```

```
end.map do |odd_number|
    odd_number**3
end
```

```
=> [1, 27, 125, 343, 729, 1331, 2197, 3375, 4913, 6859]
```

This new array is clearly a list of odd cubes. (Do the math by hand for a few of them if you don't believe me). Now, all we have to do is sum them up (or reduce the list by adding up the elements). The initial value of the sum is 0. For example:

```
(1..20).select do |number|
    number%2 == 1
end.map do |odd_number|
    odd_number**3
end.reduce(0) do |previous_sum, odd_cube|
    previous_sum + odd_cube
end
```

```
=> 19900
```

Now, I'll just change the range of numbers to 1000:

```
(1..1000).select do |number|
    number%2 == 1
end.map do |odd_number|
    odd_number**3
end.reduce(0) do |previous_sum, odd_cube|
    previous_sum + odd_cube
end
```

```
=> 124999750000
```

This answer is concise, but, more importantly, we were able to build it up by staging the data processing with select, map, and reduce operations working in a cascade.

Sort a List of Names by Surname

Suppose you have a list of names in a simple text file called names.txt.

```
John Smith
Azamat Bagatov
Hafaz Aladeen
Ramachandran Balasubrahmaniam
Ping Li
Wilfredo Caguiat
Eriks Ivanans
Canaan Banana
Ion Iliescu
Werner Klempner
Thierry Giscard
Joao Soares
```

First, you read the file into a giant string.

```
names = File.open("names.txt","r").read
```

You can split the string into an array like this:

```
namelist = names.split("\n")
```

That yields an array that looks like this:

```
> namelist
```

```
["John Smith", "Azamat Bagatov", "Hafaz Aladeen", "Ramachandran
Balasubrahmaniam\
","Ping Li","Wilfredo Caguiat","Eriks Ivanans","Canaan
Banana","Ion Iliescu","We\
rner Klempner", "Thierry Giscard","Joao Soares"]
```

The problem is that we cannot simply sort the list by string, because we need to sort by surname. Let us see if we can split off the last names and sort them. First, we need to split up the names using the split function on strings:

```
namelist.map do |fullname|
    fullname.split(" ")
end
```

This yields the following:

```
[["John", "Smith"], ["Azamat", "Bagatov"], ["Hafaz","Aladeen"]
# and so on]
```

From there, we can just peel off the last name by taking the [1] element of the name array:

```
namelist.map do |fullname|
    fullname.split(" ")
end.map do |namearray|
    namearray[1]
end
```

Now, we have a list of surnames as follows: ["Smith", "Bagatov", "Aladeen" # and so on]

We can sort these by calling the sort function:

```
namelist.map do |fullname|
    fullname.split(" ")
end.map do |namearray|
    namearray[1]
end.sort
```

Now, we have a list that looks like this:

```
["Aladeen", "Bagatov", "Balasubrahmaniam" #and so on ...
]
```

34

The challenge is to get the full names printed in order of the surname. One way is to start by annotating the array with the sorting key. The sorting key is just the second element of the array, which is yielded by splitting the full name. In other words:

```
fullname.split(" ")[1].
namelist.map do |fullname|
    [fullname.split(" ")[1], fullname]
end
```

Here, the first call to map yields array elements that look like this:

```
[ ["Smith", "John Smith"], ["Bagatov"], "Azamat Bagatov"] #...
and so on]
```

Now, we just have to sort the array based on the first element of the array:

```
namelist.map do |fullname|
    [fullname.split(" "), fullname]
end.sort do |a,b|
    a[0] <=> b[0]
end
```

That yields something that looks like this:

```
[["Aladeen", "Hafaz Aladeen"], ["Bagatov", "Azamat Bagatov"],
["Balasubrahmaniam\
", "Ramachandran Balasubrahmaniam"] #...and so on
]
```

After this operation, it only remains to strip off the sorting key to get the original list, and then to reconstitute the list.

```
namelist.map do |fullname|
    [fullname.split(" "), fullname]
```

```
end.sort do |a,b|
    a[0] <=> b[0]
end.map do |key_with_name|
    key_with_name[1]
end.join("\n")
```

The full data-processing operation then looks like this:

```
File.open("names.txt","r")
    .read
    .split("\n")
    .map do |fullname|
        fullname.split(" ")
    end.map do |fullname|
        [fullname.split(" ")[1], fullname]
    end.sort do |a,b|
        a[0] <=> b[0]
    end.map do |key_with_name|
        key_with_name[1]
    end.join("\n")
```

This is a pretty elegant solution. This could be coded with `for` loops and `if-then-else` statements, but it would result in a lot more code.

Convert a List of Names to CSV

One typical problem is to get a text file that you need to be able to import into a CSV file. For example, in `names.txt` you may have the following:

```
John Smith
Azamat Bagatov
Hafaz Aladeen
```

```
Ramachandran Balasubrahmaniam
Ping Li
Wilfredo Caguiat
```

But to import it into CSV, you need something that looks like this:

```
Surname,First Name

Smith,John
Bagatov,Azamat
Aladeen,Hafaz
Balasubrahmaniam,Ramachandran
Li,Ping
Caguiat,Wilfredo
```

The first thing you want to do is read the file:

```
file_contents = File.open("names.txt","r")
```

The next thing you want to do is split the lines into an array:

```
names = file_contents.split("\n")
```

That yields something like this:

```
["John Smith", "Azamat Bagatov" # etc
]
```

For each name, you want to split the first and last names and reverse them:

```
"John Smith".split(" ")
```

That yields the following:

```
["John","Smith"]
```

Reverse it as follows:

```
"John Smith".split(" ").reverse
```

This yields the following:

```
["Smith","John"]
```

To make it a good CSV row, you need to join the elements with a comma:

```
["Smith","John"].join(",")
```

That yields the following:

```
"Smith,John"
```

Now, we have to do that to each line of the list:

```
headerless_names =
names.map do |name|
  name.split(" ").reverse.join(",")
end
```

That yields everything except the headers. We can just use the unshift() method on the array to add the header information:

```
headered_names
headerless_names.unshift("Surname,First Name","")
```

To join the array into a string writable to a CSV file, do the following:

```
headered_names.join("\n")
```

This yields the desired result. The full computation, then, is as follows:

```
csv_file_contents =
File.open("names.txt","r").split("\n").map do |name|
  name.split(" ").reverse.join(",")
end.unshift("Surname,First  Name"),"").join("\n")
```

Generate a Random List of Names

You have some data processing to do, and you need a large list of names. For example:

```
John Smith
Azamat Bagatov
Hafaz Aladeen
Tamir Moufrad
```

However, you don't want to spend a day writing them. Use a computer to do it for you.

Start with a list of first names in first.txt. Make this list as long as you like. The longer the better.

```
Art
Bart
Cart
Dart
Eart
Fred
George
Harry
Ian
John
Ken
Lionel
Mark
```

Start with a list of surnames in last.txt:

```
Armey
Burr
Catherwood
Donaldson
```

```
Elwood
Firestone
Graves
Harvey
Illington
Jask
Knowles
Lescon
Merriweather
```

First, you need to read in the names:

```
last_name_file_content = File.open("last.txt")
first_name_file_content = File.open("first.txt")
```

Next, you need to split them into an array:

```
last_names = last_name_file_content.split("\n")
first_names = first_name_file_content.split("\n")
```

Then, you need to join every first name with every last name. Try joining each of the first names with the first last name:

```
test_list = first_names.map do |first_name|
    "#{first_name} Armey"
end
```

So far, so good. That yields an array that looks like this:

```
["Art Armey", "Bart Armey", "Cart Armey" ....and so on
```

Now, you just have to do this for all the last names:

```
test2_list =
last_names.map do |last_name|
  first_names.map do |first_name|
```

```
    "#{first_name} #{last_name}"
  end
end
```

That gives us a list that looks like the following:

```
[["Art Armey", "Bart Armey", ... ],["Art Burr", "Bart Burr",
....]]
```

That's close. We just need to flatten the array:

```
test2_list.flatten
```

That gives us the list of people. After that, we just have to join it to get the text to write to a file:

```
test2_list.flatten.join("\n")
```

The complete computation, then, is as follows:

```
last_names = File.open("last.txt").split("\n")
first_names = File.open("first.txt").split("\n")

name_list_text =
last_names.map do |last_name|
  first_names.map do |first_name|
    "#{first_name} #{last_name}"
  end
end.flatten.join("\n")
```

The beauty of this is that for the "price" of typing out 32 first names and 32 last names, we can generate 1,024 unique names.

Clean a Data Set

A common problem when analyzing data sets is dirty data. For example, you have a CSV list of names and provinces in a file called name_prov.txt.

```
John,Ontario
Azamat,ON
Hafaz,Alta
Tamir,QC
Jacques,Quebec
Bill,AB
Bart,Ont
Carn,Que
Dave,Alta
Don,SK
Fred,Saskatchewan
Gary,Que
Hari,Sask
Ian,SK
Jerry,Alberta
Karl,ON
```

A Canadian reading this list will often know that any of "ON," "Ont," and "Ontario" mean Ontario, that "AB," "Alta," and "Alberta" mean Alberta, and so on for other provinces. One way of dealing with this is to clean the data set and standardize the province names. The data transformation takes any of "ON," "Ont," and "Ontario" and replaces it with the standard name "Ontario." (As the data cleaner, you could pick any one. I have chosen to go with the full name).

The first step is to get a list of provinces:

```
province_file = File.open("name_prov.txt").read.split(("\n")
```

This yields an array of rows. Now, we will split each row on the comma and just take the second one (index 1 in the array)

```
provinces = province_file.map do |row|
    row.split(",")[1]
end

=> ["Ontario","ON","Alta","QC","Quebec","AB","Ont","Que","Alta",
"SK","Saskatchew\
an","Que","Sask","SK","Alberta","ON"]
```

This yields an array of province names, but some names are duplicated. We need distinct names. To do this, we apply the array method uniq():

```
provinces.uniq
```

This gives us an unduplicated list of province labels.

Now, we decide to encode the standardizing rules. We'll do it in a Hash to look up the name and return the standard name.

```
standard_name_lookup =
    { "ON" => "Ontario",
      "Ont" => "Ontario",
      "Ontario" => "Ontario",
      "Alta" => "Alberta",
      "Alberta" => "Alberta",
      "AB" => "Alberta",
      "QC" => "Quebec",
      "Que" => "Quebec",
      "Quebec" => "Quebec",
      "SK" => "Saskatchewan",
      "Sask" => "Saskatchewan",
      "Saskatchewan" => "Saskatchewan"
    }
```

Now, if we want to look up the province called "AB," we just do this:

```
standard_name_lookup["AB"]
```

```
=> "Alberta"
```

We'll want to encode those rules in a CSV file called standard_name_lookup.txt, which looks like this:

```
ON,Ontario
Ont,Ontario
Ontario,Ontario
Alta,Alberta
Alberta,Alberta
AB,Alberta
QC,Quebec
Que,Quebec
Quebec,Quebec
SK,Saskatchewan
Sask,Saskatchewan
Saskatchewan,Saskatchewan
```

This enables the lookup policy data (encoded in standard_name_lookup.txt) to be separate from the source code. That way, you only have to edit *this* file if you wish to change the lookup policy—without changing the software.

Reading the file and converting it to a Hash then becomes:

```
standard_name_lookup =
File.open("standard_name_lookup.txt","r").read.split("\n").map
do |row|
    row.split(",")
end.reduce({}) do |lookup_table, row|
        lookup_table.merge row[0] => row[1]
end
```

So, the only job that remains is substituting the varied province labels with the standard one:

```
standardized_province_file =
province_file.map do |row|
    row.split(",")
end.map do |split_row|
    [split_row[0], standard_name_lookup[split_row[1]]]
end
```

This yields an array with the names replaced. From there only has to join each row element with a comma and then join each row with a carriage return. The full code then becomes as follows:

```
standard_name_lookup =
File.open("standard_name_lookup.txt","r").read.split("\n").map
do |row|
    row.split(",")
end.reduce({}) do |lookup_table, row|
        lookup_table.merge row[0] => row[1]
end

standardized_province_file =
province_file.map do |row|
    row.split(",")
end.map do |split_row|
    [split_row[0], standard_name_lookup[split_row[1]]]
end.map do |standardized_split_row|
    standardized_split_row.join(",")
end.join("\n")
```

Annotate a Sequence of Sales with Running Total

Start with a sequence of numbers—say, representing the dollar value of sales.

```
sales_data = [1,2,4,7,9,11]
```

Annotate each number with the current running total. The final list should look like this:

```
[[1,1],[2,3],[4,7],[7,14],[9,23],[11,34]]
```

At first glance, this looks like an application for map(), because we are annotating each sales figure with a number. However, the number is a cumulative total of previous sales figures, which intuitively suggests using reduce().

For example, one of the following equivalent operations:

```
[1,2,4,7,9,11].reduce(0,:+)
```

or

```
[1,2,4,7,9,11].reduce(0){|run_total, sales_fig| run_total + sales_fig}
```

or

```
[1,2,4,7,9,11].reduce(0) do |running_total, sales_figure|
    running_total + sales_figure
end
```

This yields a total of 34.

Also, we know that our result will be an array of arrays, and our initial value will be a blank array. Also, each iteration will have to push an array of two elements into the initial array. The first element of the inner array holds the current sales figure, while the second one holds the running total. Try the following to get the structure right:

```
test1 =
sales_data.reduce([]) do |memo, sales_figure|
  memo.push([sales_figure, 666])
end
```

This yields the following:

```
[[1, 666], [2, 666], [4, 666], [7, 666], [9, 666], [11, 666]]
```

We're on the right track, but we now have to figure out how to get the running total into the second element of the inner arrays. The running total, of course, is the last cumulative total plus the current sales figure.

```
test2 =
sales_data.reduce([]) do |memo, sales_figure|
  memo.push([sales_figure, memo.last[-1] + sales_figure])
end
```

This throws up an exception because memo.last has nothing in it on the first run of the iteration. We need to handle that case by assigning it a 0, as follows:

```
test2 =
sales_data.reduce([]) do |memo, sales_figure|
  last_total = memo.last.nil? ? 0 : memo.last[1]
  memo.push([sales_figure, last_total + sales_figure])
end
```

This yields the correct answer:

```
[[1, 1], [2, 3], [4, 7], [7, 14], [9, 23], [11, 34]]
```

Pascal's Triangle

A classic problem in computer science is the one of generating Pascal's triangle. For example, a six-row triangle looks like this:

```
1
1  1
1  2  1
1  3  3  1
1  4  6  4  1
1  5 10 10  5  1
1  6 15 20 15  6  1
```

In general, if one stores it as a two-dimensional array, the formula is as follows:

```
a[n][0] = 1    # where n >=0
a[n][n] = 1
a[n][m] = a[n-1][m] + a[n-1][m-1]
```

First, this is obviously a two-dimensional array. So, it will look something like this for a four-row triangle:

```
[[1],[1,1],[1,2,1],[1,3,3,1]]
```

For convenient printing, we'll just do this for now:

```
triangle.each{|row| p row.inspect}
```

Given that we're building a structure of arrays, it seems like a good fit for using reduce(). The initial value is [].

```
triangle = (0..5).reduce([]) do |memo, n|
  memo.push []
end

triangle.each{|row| p row.inspect}

"[]"
"[]"
"[]"
"[]"
"[]"
"[]"
=> [[], [], [], [], [], []]
```

That yields an array of arrays, but not enough is contained in it. However, we know that the first and last values are 1. We can default the rest of the values to zero. And we know that each array has *n*-1 elements. (Note that row[-1] is the way to denote the last element of an array).

```
triangle = (0..5).reduce([]) do |memo, n|
  row = n==0 ? [] : Array.new(n+1)
  row[0] = 1
  row[-1] = 1
  memo << row
end
=> [[1], [1, 1], [1, nil, 1], [1, nil, nil, 1], [1, nil, nil,
nil, 1], [1, nil, \
nil, nil, nil, 1]]
```

The structure is taking shape, but we also know that the *n*th row is a function of the previous row. So, we need to dig back to the last row and muck with it to generate the next row.

```
triangle = (0..5).reduce([]) do |memo, n|
  if n==0
    row = [1]
  else
    prev = acc [-1]
    row = prev << 2
  end
  memo << row
end

=> [[1], [1, 1], [1, nil, 1], [1, nil, nil, 1], [1, nil, nil,
nil, 1], [1, nil, \
nil, nil, nil, 1]]
```

Or, better yet:

```
irb(main):007:0> triangle.each{|row| p row.inspect}
"[1]"
"[1, 1]"
"[1, nil, 1]"
"[1, nil, nil, 1]"
"[1, nil, nil, nil, 1]"
"[1, nil, nil, nil, nil, 1]"
=> [[1], [1, 1], [1, nil, 1], [1, nil, nil, 1], [1, nil, nil,
nil, 1], [1, nil, \
nil, nil, nil, 1]]
```

This is a little better. The rows are starting to flesh out. Now, to build the current row, you need the previous row.

For example, you could reduce the previous row to get the current row. To get the last two elements of a row, you need to remember them. For example:

```
[1,3,3,1].reduce([nil,nil]) do |memo, element|
    memo.push element
    memo.shift
    puts "After pushing #{element}, the memo is #{memo.inspect}"
    memo
end
```

```
After pushing 1, the memo is [nil, 1]
After pushing 3, the memo is [1, 3]
After pushing 3, the memo is [3, 3]
After pushing 1, the memo is [3, 1]
=> [3, 1]
```

We're getting close. Let's save the sum in the memo. Now, remember that the structure of the memo must change. We need a place to save the last two elements in an array, as well as the sum. So, the memo will be a Hash, with :sum and :last_2 as keys.

```
initial_memo = {sum: 0, last_2: [nil,nil]}
[1,3,3,1].reduce(initial_memo) do |memo, element|
    memo[:last_2].push element
    memo[:last_2].shift
    memo[:sum] = memo[:last_2].compact.reduce(0,:+)
    puts "After pushing element #{element}, memo is #{memo}"
    memo
end
```

The result is as follows:

```
After pushing element 1, memo is {:sum=>1, :last_2=>[nil, 1]}
After pushing element 3, memo is {:sum=>4, :last_2=>[1, 3]}
After pushing element 3, memo is {:sum=>6, :last_2=>[3, 3]}
After pushing element 1, memo is {:sum=>4, :last_2=>[3, 1]}
=> {:sum=>4, :last_2=>[3, 1]}
```

We're even closer. The sum follows the pattern 1,4,6,4, which is almost the next row (missing the last 1). So, now we need to save the next row in the memo and then extract it when the computation is done, as follows:

```
initial_memo = {last_2: [nil,nil], next_row: []}
[1,3,3,1].reduce(initial_memo) do |memo, element|
    memo[:last_2].push element
    memo[:last_2].shift
    memo[:next_row].push memo[:last_2].compact.reduce(0,:+)
    puts "After pushing element #{element}, memo is #{memo}"
    memo
end
```

The result is as follows:

```
After pushing element 1, memo is {:last_2=>[nil, 1], :next_
row=>[1]}
After pushing element 3, memo is {:last_2=>[1, 3], :next_
row=>[1, 4]}
After pushing element 3, memo is {:last_2=>[3, 3], :next_
row=>[1, 4, 6]}
After pushing element 1, memo is {:last_2=>[3, 1], :next_
row=>[1, 4, 6, 4]}
=> {:last_2=>[3, 1], :next_row=>[1, 4, 6, 4]}
```

Almost! We just have to extract the :next_row from the result and push a 1 to the end of the next row. I'll also remove the puts statement.

```
initial_memo = {last_2: [nil,nil], next_row: []}
[1,3,3,1].reduce(initial_memo) do |memo, element|
    memo[:last_2].push element
    memo[:last_2].shift
    memo[:next_row].push memo[:last_2].compact.reduce(0,:+)
    memo
end[:next_row].push 1
```

=> [1, 4, 6, 4, 1]

So, we have a formula for each row. Let's see if it works for a few examples. First, the top row:

```
initial_memo = {last_2: [nil,nil], next_row: []}
[1].reduce(initial_memo) do |memo, element|
    memo[:last_2].push element
    memo[:last_2].shift
    memo[:next_row].push memo[:last_2].compact.reduce(0,:+)
    memo
end[:next_row].push 1
```

=> [1, 1]

How about a blank row (for example, [ ])?

```
initial_memo = {last_2: [nil,nil], next_row: []}
[].reduce(initial_memo) do |memo, element|
    memo[:last_2].push element
    memo[:last_2].shift
    memo[:next_row].push memo[:last_2].compact.reduce(0,:+)
    memo
end[:next_row].push 1
```

=> [1]

This seems to work. What remains is to do this for every row. Again, we'll use reduce because each iteration carries a memo to the next one. Try this (say, for seven rows):

```
inital_memo = [[]]
(1..7).reduce([[]]) do |memo, num|
        initial_memo = {last_2: [nil,nil], next_row: []}
        memo.push(
            memo[-1].reduce(initial_memo) do |memo, element|
                memo[:last_2].push element
                memo[:last_2].shift
                memo[:next_row].push memo[:last_2].compact.
                reduce(0,:+)
                memo
            end[:next_row].push(1)
        )
        puts "After processing #{num}, memo is #{memo}"
        memo
end
```

This works. What remains is to get rid of the puts statements and print it out properly.

```
inital_memo = [[]]
triangle =
(1..7).reduce([[]]) do |memo, num|
        initial_memo = {last_2: [nil,nil], next_row: []}
        memo.push(
            memo[-1].reduce(initial_memo) do |memo, element|
                memo[:last_2].push element
                memo[:last_2].shift
                memo[:next_row].push memo[:last_2].compact.
                reduce(0,:+)
                memo
```

```ruby
        end[:next_row].push(1)
      )
      puts "After processing #{num}, memo is #{memo}"
      memo
end
triangle.each{|row| p row.inspect}
"[]"
"[1]"
"[1, 1]"
"[1, 2, 1]"
"[1, 3, 3, 1]"
"[1, 4, 6, 4, 1]"
"[1, 5, 10, 10, 5, 1]"
"[1, 6, 15, 20, 15, 6, 1]"
```

A cleaner version would be as follows:

```ruby
triangle.each{|row| puts row.join (", ") };1

1
1, 1
1, 2, 1
1, 3, 3, 1
1, 4, 6, 4, 1
1, 5, 10, 10, 5, 1
1, 6, 15, 20, 15, 6, 1
```

Reverse Polish Notation Parser

One way to describe arithmetic operations is to use notation where the operands for an operation come before the operator. For example, the expression

```ruby
3 + 2
```

would be written as

3 2 +

This is called *postfix* or *reverse polish* notation. (It was a big feature of the Hewlett-Packard calculators used by engineering geeks back in the 1980s when I went to school). It lets you avoid using parentheses. Processing is easier because one only needs to use a stack. As each token (number or operator) is read, it is pushed onto the stack. When an operator is pushed on the stack, it is a special case. The operator is popped off the stack along with the next two elements, which are the operands, and the operation is performed. The result is pushed back on the stack.

So, for example, the processing of the expression

3 3 4 * 4 7 8 + - * +

can be shown as a sequence of stack snapshots, one per line. The stack grows to the right as the tokens are scanned, and if an operator is pushed on the stack, the RPN interpreter pops off the last three tokens, runs the calculation, and pushes the answer back onto the stack.

```
3
3 3
3 3 4
3 3 4 *
3 12                    (12 is 4*3)
3 12 4
3 12 4 7
3 12 4 7 8
3 12 4 7 8 +
3 12 4 15               (15 is 7+8)
3 12 4 15 -
3 12 -11                (-11 is 4-15)
```

```
3  12  -11  *
3  -132                  (-132 is 12*(-11))
3  -132  +
-129                     (-129 is 3 + (-132))
```

So, how to solve this? I'll start by saying that we want to tokenize a string by splitting it by a space. (This is not the most robust way to tokenize the string, but it's good enough for this example). Once we have an array of tokens from the string, we will probably want to remember the stack after each iteration. Now, a Ruby array can be used as a stack by just using its push and pop operations. Also, the initial value of the stack is the null array. Let's try the following:

```ruby
expression = "1 2 2 + 3 * -"
answer = expression.split(" ").reduce([]) do |stack, token|
    stack.push(token)
    puts "The stack after pushing #{token} is #{stack.inspect}"
    stack
end
```

This prints out the following:

```
The stack after pushing 1 is ["1"]
The stack after pushing 2 is ["1", "2"]
The stack after pushing 2 is ["1", "2", "2"]
The stack after pushing + is ["1", "2", "2", "+"]
The stack after pushing 3 is ["1", "2", "2", "+", "3"]
The stack after pushing * is ["1", "2", "2", "+", "3", "*"]
The stack after pushing - is ["1", "2", "2", "+", "3", "*", "-"]
=> ["1", "2", "2", "+", "3", "*", "-"]
```

So, we're saving the stack properly after each iteration. Now, we need to identify the operators "+", "-", "*", and "/". Let's try this:

```
expression = "1 2 2 + 3 * -"
answer = expression.split(" ").reduce([]) do |stack, token|
    stack.push(token)
    puts "The stack after pushing #{token} is #{stack.inspect}"
    if ["+","-","*","/"].include?(token)
        puts " Token #{token} is an operator"
    end
    stack
end
```

That prints out the following:

```
The stack after pushing 1 is ["1"]
The stack after pushing 2 is ["1", "2"]
The stack after pushing 2 is ["1", "2", "2"]
The stack after pushing + is ["1", "2", "2", "+"]
    Token + is an operator
The stack after pushing 3 is ["1", "2", "2", "+", "3"]
The stack after pushing * is ["1", "2", "2", "+", "3", "*"]
    Token * is an operator
The stack after pushing - is ["1", "2", "2", "+", "3", "*", "-"]
    Token - is an operator
=> ["1", "2", "2", "+", "3", "*", "-"]
```

Now, as per RPN, we want to do an operation if the token is an operator. To do the operation, we have to pop the operator off the stack, then pop the first operand, and finally pop the second operand. I'm going to stop this iteration the first time we get an operator just to see what the stack looks like.

```ruby
expression = "1 2 2 + 3 * -"
answer = expression.split(" ").reduce([]) do |stack, token|
    stack.push(token)
    puts "The stack after pushing #{token} is #{stack.inspect}"
    if ["+","-","*","/"].include?(token)
        puts " Token #{token} is an operator"
        operator = stack.pop
        second_operand = stack.pop
        first_operand = stack.pop
        puts " Second operand is #{second_operand}. First
        operand is #{first_operand}"
        raise "Stop for a bit"
    end
    stack
end
```

This prints out the following:

```
The stack after pushing 1 is ["1"]
The stack after pushing 2 is ["1", "2"]
The stack after pushing 2 is ["1", "2", "2"]
The stack after pushing + is ["1", "2", "2", "+"]
    Token + is an operator
    Second operand is 2. First operand is 2
Stop for a bit
(repl):11:in `block in initialize'
(repl):2:in `each'
(repl):2:in `reduce'
(repl):2:in `initialize'
```

The Stop for a bit exception only lets us get to the first operator.
I just wanted to show the operator and operands getting popped off
the stack. Now, we just have to send the "operator" message to the first

operand, with the second operand as a parameter. (Remember that both 5 and 7 are Ruby objects of class Fixnum, and they both have a method called :+). For example,

```
5 + 7
```

is the same as saying

```
5.send(:+,7)
```

Try it out on the command line if you don't believe me.

Now, each token popped off the stack is a string. The operands must be converted to integers (using to_i), and the operator must be converted to a symbol (using to_sym). So, the code snippet would look like this:

```
first_operand.to_i.send(operator.to_sym, second_operand.to_i)
```

This tells the interpreter to call the :+ method (because "+".to_sym is :+) of the 5 first_operand object, which happens to be a Fixnum. Once that result is calculated, the result must be pushed back on the stack. (The stack variable is the memo of this reduce operation).

```
expression = "1 2 2 + 3 * -"
answer = expression.split(" ").reduce([]) do |stack, token|
    stack.push(token)
    puts "The stack after pushing #{token} is #{stack.inspect}"
    if ["+","-","*","/"].include?(token)
        puts " Token #{token} is an operator"
        operator = stack.pop
        second_operand = stack.pop
        first_operand = stack.pop
        puts " Second operand is #{second_operand}. First
        operand is #{first_operand}"
```

```
    result = first_operand.to_i.send(operator.to_sym,
    second_operand.to_i)
    puts " The result of this expression is #{result}"
    stack.push(result)
    puts " The stack after evaluating the expression is
    #{stack.inspect}"
  end
  stack
end
```

This prints out the following:

```
The stack after pushing 1 is ["1"]
The stack after pushing 2 is ["1", "2"]
The stack after pushing 2 is ["1", "2", "2"]
The stack after pushing + is ["1", "2", "2", "+"]
  Token + is an operator
  Second operand is 2. First operand is 2
  The result of this expression is 4
    The stack after evaluating the expression is ["1", 4]
The stack after pushing 3 is ["1", 4, "3"]
The stack after pushing * is ["1", 4, "3", "*"]
  Token * is an operator
  Second operand is 3. First operand is 4
  The result of this expression is 12
    The stack after evaluating the expression is ["1", 12]
The stack after pushing - is ["1", 12, "-"]
  Token - is an operator
  Second operand is 12. First operand is 1
  The result of this expression is -11
    The stack after evaluating the expression is [-11]
```

Now, to clean up the code and print out the answer, we do the following:

```
expression = "1 2 2 + 3 * -"
answer = expression.split(" ").reduce([]) do |stack, token|
    stack.push(token)
    if ["+","-","*","/"].include?(token)
        operator = stack.pop
        second_operand = stack.pop
        first_operand = stack.pop
        result = first_operand.to_i.send(operator.to_sym,
        second_operand.to_i)
        stack.push(result)
    end
    stack
end[0]
puts "The expression is #{expression}. The answer is #{answer}."
```

Generate a List of Team Name Bars

Suppose you have a list of names in a simple text file called names.txt. These are the names of the girls on a basketball team.

```
Emily Muller
Lois Makad
Sophie Wheaton
Abby Muller
Zoe Gee
Anastasia Semanova
Iulia Semanova
Jane Smith
Megan Sundstrom
```

Brunhilda Schmidt
Sally Park
Jill Bailey

You want to generate a list of name bars to put on the player jerseys, using only the surnames, unless there are two people with the same surname. In that case, you want the name bar to have the first initial and the surname. The list, then, would look like this:

E. Muller
Makad
Wheaton
A. Muller
Gee
A. Semanova
I. Semanova
Smith
Sundstrom
Schmidt
Park
Bailey

First, you read the file into a giant string:

```
names = File.open("names.txt","r").read
```

You can split the string into an array like this:

```
namelist = names.split("\n")
```

That yields an array that looks like this:

```
["Emily Muller", "Lois Makad", "Sophie Wheaton", # and so on
]
```

First, you want to annotate the list with the surname to help:

```ruby
annotated_namelist = namelist.map do |name|
  [name.split(" ")[1], name]
end
```

Then, you want to group the names by distinct surname. A Ruby hash is a useful way to key by surname:

```ruby
grouped_list =
annotated_namelist.reduce({}) do |memo, annotated_name|
    if memo[annotated_name[0]].nil?
        memo[annotated_name[0]] = [annotated_name[1]]
    else
        memo[annotated_name[0]] << annotated_name[1]
    end
    memo
end
```

At this point, we have hash-keyed by distinct surname, and it looks like this:

```ruby
{"Muller"    =>   ["Emily Muller", "Abby Muller"],
 "Makad"     =>   ["Lois Makad"],
 "Wheaton"   =>   ["Sophie Wheaton"],
 "Gee"       =>   ["Zoe Gee"],
 "Semanova"  =>   ["Anastasia Semanova", "Iulia Semanova"],
 "Smith"     =>   ["Jane Smith"]
 "Sundstrom" =>   ["Megan Sundstrom"],
 "Schmidt"   =>   ["Brunhilda Schmidt"],
 "Park"      =>   ["Sally Park"],
 "Bailey"    =>   ["Jill Bailey"]
}
```

We should note that the key is a string and the value is an array of name(s). We need to reduce the name to the name bar. If there is one element (name) in the array, we just have to split the string and take the second element, which is the surname. If there is more than one element, we need to split the string, take the first letter of the first element, and concatenate it with the second element. See here:

```
grouped_label_names =
grouped_list.map do |key,value|
  if value.size>1
    res = value.map do |name|
      "#{name.split(' ')[0][0]}. #{name.split(' ')[1]}"
    end
    res
  else
    value[0].split(' ')[1]
  end
end
```

This yields the following:

```
[["E. Muller", "A. Muller"], "Makad", "Wheaton", "Gee",
["A. Semanova", "I. Sema\
nova"], "Smith", "Sundstrom", "Schmidt", "Park", "Bailey"]
```

This is almost good. The list just has to be flattened:

```
grouped_label_names.flatten
```

This yields the following:

```
["E. Muller", "A. Muller", "Makad", "Wheaton", "Gee",
"A. Semanova", "I. Semanov\
a", "Smith", "Sundstrom", "Schmidt", "Park", "Bailey"]
```

The list can then be reconstituted simply by joining with a carriage return, as follows:

```
grouped_label_names.flatten.join("\n")
```

Once this is done, the array has to be flattened and reconstituted back into the text file:

```
namebar_names =
grouped_label_names.flatten.join("\n")
```

Thus, the complete data-processing solution is as follows:

```
namebar_names =
  File.open("names.txt","r").read.split("\n").map do |name|
          [name.split(" ")[1], name]
      end.inject({}) do |memo, annotated_name|
        if memo[annotated_name[0]].nil?
          memo[annotated_name[0]] = [annotated_name[1]]
        else
          memo[annotated_name[0]] << annotated_name[1]
        end
        memo
      end.map do |key,value|
        if value.size>1
          res = value.map do |name|
            "#{name.split(' ')[0][0]}. #{name.split(' ')[1]}"
          end
          res
        else
          value[0].split(' ')[1]
        end
      end.flatten.join("\n")
```

Find Intersection Points of Two Functions

Suppose you have two functions, $y = x$ and $y = 10sin(x)$. You want to find their intersection points. A few facts:

- $y = 10sin(x)$ has a period of $2*pi$, or about 6.2, so there is an actual zero every half period, or about every 3.14.

- $y = 10sin(x)$ has a maximum value of 10 and a minimum value of -10, so if we are going to search the line $*y = x*$, we can limit x to also be between -10 and 10.

- From calculus, we know the slope of $y = 10sin(x)$ is $y = 10cos(x)$, and it takes a maximum value of 10 when $x=0$, as well as at other places.

- For a slope of 10, if you want to change y by up to some small amount, the change in x has to be at most a tenth of that amount.

- Intersection occurs when the difference between $y=10sin(x)$ and $y = x$ is 0. But, in an imperfect world, we could say that we want the difference to be less than, say, 0.02.

- To guarantee the difference, we must test for intersection in steps 1/10 the size, or 0.002. Thus, the domain (in other words, the set of x values to test) can be expressed as follows:

  (-10..10).step(0.002)

Print it out by doing this:

```
(-10..10).step(0.002).map{|n| n}
```

That spits out 2000 values of *x* to test. For each of those values, you want to collect the range for both functions. Each of these is an array of three elements comprising the *x* value, the *y* value of the first function (*y=x*), and the *y* value of the second function (*y=10sin(x)*). See here:

```
function_values =
(-10..10).step(0.002).map do |x|
    [x, x, 10*Math.sin(x)]
end
```

Now, the intersection of these two functions occurs when the difference in the two functions (for example, element 1, and 2 in the three-tuple) is extremely small. In our example, we stepped the *x* value by 0.002 every time, so the maximum change we would expect is 0.02. So, a difference smaller than 0.02 can be practically considered a zero. Out of all two-thousand-ish of those three-tuples, we only want the ones where elements 1 and 2 (not 0) are very close. In other words, the absolute value of the difference should be less than 0.02. See the following:

```
zeroes = function_values.select do |three_tuple|
    (three_tuple[0] - three_tuple[1]).abs < 0.02
end
```

That yields a set of points in both functions that are small enough to be considered zeroes.

```
zeroes =
(-10..10).step(0.002).map do |x|
    [x, x, 10*Math.sin(x)]
end.select do |three_tuple|
    (three_tuple[1] - three_tuple[2]).abs < 0.02
end
```

This yields the following (which I have formatted for easy reading):

```
=> [
    [-8.426, -8.426, -8.408100860402287],
    [-8.424, -8.424, -8.418910641243016],
    [-8.422, -8.422, -8.429686746452397],
    [-7.07, -7.07, -7.081077089878837],
    [-7.068, -7.068, -7.066940848453371],
    [-7.066, -7.066, -7.052776339273938],
    [-2.8539999999999996, -2.8539999999999996,
    -2.836445737876892],
    [-2.852, -2.852, -2.855618641886548],
    [-0.001999999999999792, -0.001999999999999792,
    -0.01999998666666725],
    [2.0816681711721685e-16, 2.0816681711721685e-16,
    2.0816681711721685e-15],
    [0.002000000000000208, 0.002000000000000208,
    0.019999986666671413],
    [2.8520000000000003, 2.8520000000000003,
    2.8556186418865437],
    [2.854, 2.854, 2.8364457378768875],
    [7.066000000000001, 7.066000000000001, 7.0527763392739455],
    [7.0680000000000005, 7.0680000000000005, 7.066940848453377],
    [7.07, 7.07, 7.081077089878837], [8.422, 8.422,
    8.429686746452397],
    [8.424000000000001, 8.424000000000001,
    8.418910641243006], [8.426, 8.426,
    8\.408100860402287]
]
```

69

These results are interesting because there are 19 zeroes in this list (admittedly down a lot from 2000 possible points). However, if you look carefully, they are clustered around -8.41, -7.07, -2.84, 0, 2.84, 7.07, and 8.41. The question is whether these clusters happen because there are actually that many zeroes or because the two functions are close enough (for instance, by 0.02 or less) without touching for more than one step (0.02). One way to answer this question is to collect information to see if the functions cross. To do that, we have to gather the difference in functions in the map() step, as follows:

```
intersection_points =
(-10..10).step(0.002).map do |x|
    [x, x, 10*Math.sin(x), x - (10*Math.sin(x))]
end.select do |four_tuple|
    four_tuple[3].abs < 0.02
end

=> [
        [-8.426, -8.426, -8.408100860402287, -0.017899139597712832],
        [-8.424, -8.424, -8.418910641243016, -0.005089358756983131],
        [-8.422, -8.422, -8.429686746452397, 0.00768674645239642],
        [-7.07, -7.07, -7.081077089878837, 0.011077089878837043],
        [-7.068, -7.068, -7.066940848453371,-0.0010591515466282786],
        [-7.066, -7.066, -7.052776339273938, -0.013223660726061404],
        [-2.8539999999999996, -2.8539999999999996, -2.836445737876892,
        -0.0175542621\
        23107733],
        [-2.852, -2.852, -2.855618641886548, 0.0036186418865482572],
        [-0.001999999999999792, -0.001999999999999792,
        -0.01999998666666725, 0.01799\
        998666666746],
        [2.0816681711721685e-16, 2.0816681711721685e-16,
        2.0816681711721685e-15, -1.\ 8735013540549517e-15],
```

```
  [0.002000000000000208, 0.002000000000000208,
  0.019999986666671413, -0.017999\
  986666671207],
  [2.8520000000000003, 2.8520000000000003,
  2.8556186418865437, -0.003618641886\
  5433723],
  [2.854, 2.854, 2.8364457378768875, 0.01755426212311262],
  [7.066000000000001, 7.066000000000001, 7.0527763392739455,
  0.013223660726055\
  186],
  [7.0680000000000005, 7.0680000000000005,
  7.066940848453377, 0.00105915154662\
  38377],
  [7.07, 7.07, 7.081077089878837, -0.011077089878837043],
  [8.422, 8.422, 8.429686746452397, -0.00768674645239642],
  [8.424000000000001, 8.424000000000001, 8.418910641243006,
  0.0050893587569955\
  66],
  [8.426, 8.426, 8.408100860402287, 0.017899139597712832]
]
```

In all cases, it appears that the functions crossed over an interval of two or three steps because the three element changed sign. At this point, we need to group the points into distinct clusters. Grouping by x value is tricky when the x values in a cluster are not exactly the same. However, we can adjust the mechanism. In general, one could argue that if two x-values of intersection points are within 0.01 of each other (for five consecutive points), they are in the same cluster. Also, this list of points above is ordered by x, so one only has to check the x value of the next point. See here:

```
intersection_clusters =
intersection_points.reduce([]) do |cluster_list, zero_four_tuple|
  if !cluster_list[-1].nil?    &&
```

```
      cluster_list[-1][0][0] - zero_four_tuple[0] > -0.01
      cluster_list[-1].push zero_four_tuple
   else
      cluster_list.push [zero_four_tuple]
   end
   cluster_list
end
```

This yields an interesting result:

```
[
   [
        [-8.426, -8.426, -8.408100860402287,
        -0.017899139597712832],
        [-8.424, -8.424, -8.418910641243016,
        -0.0050893587569831131],
        [-8.422, -8.422, -8.429686746452397,
        0.00768674645239642]
   ],
   [
        [-7.07, -7.07, -7.081077089878837,
        0.011077089878837043],
        [-7.068, -7.068, -7.066940848453371,
        -0.00105915154662827286],
        [-7.066, -7.066, -7.052776339273938,
        -0.013223660726061404]
   ],
   [
        [-2.8539999999999996, -2.8539999999999996,
        -2.836445737876892, -0.017554\
   262123107733],
```

```
    [-2.852, -2.852, -2.855618641886548,
    0.0036186418865482572]
],
[

    [-0.001999999999999792, -0.001999999999999792,
    -0.01999998666666725, 0.0\
1799998666666746],
    [2.0816681711721685e-16, 2.0816681711721685e-16,
    2.0816681711721685e-15,\
-1.8735013540549517e-15],
    [0.002000000000000208, 0.002000000000000208,
    0.019999986666671413, -0.01\
7999986666671207]
],
[

    [2.8520000000000003, 2.8520000000000003,
    2.8556186418865437, -0.00361864\
18865433723],
    [2.854, 2.854, 2.8364457378768875,
    0.01755426212311262]
],
[

    [7.066000000000001, 7.066000000000001,
    7.0527763392739455, 0.01322366072\
6055186],
    [7.0680000000000005, 7.0680000000000005,
    7.066940848453377, 0.0010591515\
466238377],
    [7.07, 7.07, 7.081077089878837,
    -0.011077089878837043]
],
```

```
[
    [8.422, 8.422, 8.429686746452397, -0.00768674645239642],
    [8.424000000000001, 8.424000000000001,
    8.418910641243006, 0.005089358756\
995566],
    [8.426, 8.426, 8.408100860402287, 0.017899139597712832]
    ]
]
```

This clearly shows seven clear clusters of zeroes. Now, at this point, you can make a decision to just accept the first zero in the cluster as "good enough."

```
intersection_clusters.map do |cluster|
    cluster[0][0]
end
```

Or, you may want the x value to be the mean of the x values in the cluster, rounded to a pretty-looking number:

```
intersection_clusters.map do |cluster|
    (cluster.reduce(0) do |sum, zero_four_tuple|
        sum + zero_four_tuple[0]
    end / cluster.size).round(4)
end
```

This yields the following:

```
=> [-8.424, -7.068, -2.853, 0.0, 2.853, 7.068, 8.424]
```

The full code is as follows:

```
zeroes =
(-10..10).step(0.002).map do |x|
    [x, x, 10*Math.sin(x), x - (10*Math.sin(x))]
end.select do |four_tuple|
```

```
    four_tuple[3].abs < 0.02
end.inject([]) do |cluster_list, zero_four_tuple|
    if !cluster_list[-1].nil?     &&
        cluster_list[-1][0][0] - zero_four_tuple[0] > -0.01
        cluster_list[-1].push zero_four_tuple
    else
        cluster_list.push [zero_four_tuple]
    end
    cluster_list
end.map do |cluster|
    (cluster.reduce(0) do |sum, zero_four_tuple|
        sum + zero_four_tuple[0]
    end / cluster.size).round(4)
end
```

Eighteen lines of code for a fairly complex calculation is pretty impressive.

Exercise: You could check for crossovers in a cluster, which happen when a cluster of tuples has differences (fourth element in the tuple) that are positive and negative.

Group by Area Code

Sometimes, you have a data set that must be grouped by a function of an element. Consider this list of phone numbers:

```
John,613-555-1234
Matt,613-555-2345
Andrew,613-555-3456
Bob,416-555-2222
Sheila,905-555-7777
Sara,416-555-9999
Grocery,416-492-9999
```

```
Pizza,416-967-1111
Gabers,613-310-7777
Shah,905-555-1212
```

You want to group the list by area code. In this case, it is the first three-digit number in each phone number.

The area code can be extracted by performing a `split()` on the phone number and taking the first element (`index 0`). For example:

```
phone_number = "613-555-1234"
area_code = phone_number.split("-")[0]

=> "613"
```

Once you have access to the area code, it can serve as a key to a `Hash` keyed by the area code. In our example:

```
area_code_groups[area_code] = phone number
```

However, we'll need an array if there are more numbers. A new array has to be created if one doesn't exist for a new area code, as follows:

```
if area_code_groups[area_code].nil?
    area_code_groups[area_code] = [phone_number]
else
    area_code_groups[area_code].push phone_number
end
```

The complete text, then, is as follows:

```
phonelist_text.split("\n").map do |row|
    row.split(",")
end.map do |split_row|
    [split_row[0], split_row[1].split("-")]
end.reduce({}) do |area_code_groups, ac_split_row|
    if area_code_groups[ac_split_row[1][0]].nil?
```

```
area_code_groups[ac_split_row[1][0]] = [ [ac_split_
  row[0], ac_split_row[1]\
.join("-")].join(",") ]
  else
    area_code_groups[ac_split_row[1][0]].push [ac_split_
      row[0], ac_split_row[1\
].join("-")].join(",")
  end
  area_code_groups
end
```

Sliding Window Average

Sometimes you have a data set that varies widely, but the long-term norm follows a set pattern. Suppose you have a list of fall temperatures in Ottawa from October 1 through October 31.

```
dailytemps = [10,20,22,23,15,18,17,10,10,10,17,20,21,18,8,16,
20,8,7,8,8,18,23,24\
,5,6,15,18,19,0,2]
```

You want to get the average of the last five days. For the first four days, just take the average of the previous four days. For the first three days, just the average of previous three days, and so on.

We can use reduce because we're doing operations that require a memo of previous calculations. Let us see if we can find the last five temperatures. The memo initializes with an array of five nils. On each iteration, we'll push the temperature onto the end of the memo and shift the oldest one off the memo. We'll only test it on the first seven elements to save space. See here:

```
initial_last_5 = [nil,nil,nil,nil,nil]
dailytemps.first(7).reduce(initial_last_5) do |memo,
temperature|
```

```
    memo.push temperature
    memo.shift
    puts "After pushing #{temperature}, the last 5 are:
    #{memo.inspect}"
    memo
end
```

```
After pushing 10, the last 5 are: [nil, nil, nil, nil, 10]
After pushing 20, the last 5 are: [nil, nil, nil, 10, 20]
After pushing 22, the last 5 are: [nil, nil, 10, 20, 22]
After pushing 23, the last 5 are: [nil, 10, 20, 22, 23]
After pushing 15, the last 5 are: [10, 20, 22, 23, 15]
After pushing 18, the last 5 are: [20, 22, 23, 15, 18]
After pushing 17, the last 5 are: [22, 23, 15, 18, 17]
=> [22, 23, 15, 18, 17]
```

We are saving the last five temperatures successfully after each step. Now, we need to save the average as well. To do that, we have to start with an initial array of the last five, as well as an initial average. A Hash is a good structure for the memo. The first key is the average, while the last key is the array of the last five elements. When we push the temperature, we need to do it to memo[:last_5] instead of just to memo. Also, the average of the last five temperatures is the sum of the non-nil elements divided by the number of non-nil elements. (compact gets rid of non-nil elements, and length is the number of non-nil elements remaining). For example:

```
example_5 = [nil,nil,6,8,13]
average = (example_5.compact.reduce(0,:+)) / (example_5.
compact.length)
```

```
=> 9
```

Nine is the correct answer for the average of the three non-nil numbers.

Now, we'll apply that to the calculation as follows:

```
initial_hash = {average: 0, last_5: [nil,nil,nil,nil,nil]}
dailytemps.first(7).reduce(initial_hash) do |memo, temperature|
    memo[:last_5].push temperature
    memo[:last_5].shift

    memo[:average] =
        (memo[:last_5].compact.reduce(0,:+)) / (Float(memo
        [:last_5].compact.leng\
th))

    puts "After pushing #{temperature}, the last 5 are: #{memo.
    inspect}"
    memo
end
```

The output looks like this:

```
After pushing 10, the memo is: {:average=>10.0, :last_5=>[nil,
nil, nil, nil, 10\
]}
After pushing 20, the memo is: {:average=>15.0, :last_5=>[nil,
nil, nil, 10, 20]}
After pushing 22, the memo is: {:average=>17.333333333333332,
:last_5=>[nil, nil\
, 10, 20, 22]}
After pushing 23, the memo is: {:average=>18.75, :last_5=>[nil,
10, 20, 22, 23]}
After pushing 15, the memo is: {:average=>18.0, :last_5=>[10,
20, 22, 23, 15]}
After pushing 18, the memo is: {:average=>19.6, :last_5=>[20,
22, 23, 15, 18]}
```

After pushing 17, the memo is: {:average=>19.0, :last_5=>[22, 23, 15, 18, 17]}
=> {:average=>19.0, :last_5=>[22, 23, 15, 18, 17]}

We're getting closer. We have the last average and the last five, but we need to save all the averages. That means that memo[:average] should be saved to an an array of sliding averages, which we push on every iteration. See here:

```
initial_hash = {average: 0,
                sliding_averages: [],
                 last_5: [nil,nil,nil,nil,nil]
                }
dailytemps.first(7).reduce(initial_hash) do |memo, temperature|
    memo[:last_5].push temperature
    memo[:last_5].shift
    memo[:average] = (memo[:last_5].compact.reduce(0,:+)) /
    (Float(memo[:last_5]\
.compact.length))
    memo[:sliding_averages].push memo[:average]
    puts "After pushing #{temperature}, the memo is: #{memo.
    inspect}"
    memo
end
```

Then, we need to keep only the sliding averages array at the end. We can get rid of the first seven and process the whole input array of temperatures, as follows:

```
initial_hash = {average: 0,
                sliding_averages: [],
                last_5: [nil,nil,nil,nil,nil]
                }
```

```
dailytemps.reduce(initial_hash) do |memo, temperature|
    memo[:last_5].push temperature
    memo[:last_5].shift
    memo[:average] = (memo[:last_5].compact.reduce(0,:+)) /
      (Float(memo[:last_5]\
.compact.length))
    memo[:sliding_averages].push memo[:average]
    puts "After pushing #{temperature}, the memo is:
    #{memo.inspect}"
    memo
end[:sliding_averages]
```

There is a whole pile of puts statements before the final answer is
returned. Inspect them to see if the outputs after each iteration make
sense. The final answer is as follows:

```
=> [10.0, 15.0, 17.333333333333332, 18.75, 18.0, 19.6, 19.0,
16.6, 14.0, 13.0, 1\
2.8, 13.4, 15.6, 17.2, 16.8, 16.6, 16.6, 14.0, 11.8, 11.8,
10.2, 9.8, 12.8, 16.2\
, 15.6, 15.2, 14.6, 13.6, 12.6, 11.6, 10.8]
```

And, to clean up the code, we don't really need to keep memo[:average]
or the puts statement. See the final version here:

```
sliding_averages =
dailytemps.reduce(sliding_averages:[],
last_5:[nil,nil,nil,nil,nil]) do |memo, t\ emperature|
    memo[:last_5].push temperature
    memo[:last_5].shift
    memo_average = (memo[:last_5].compact.reduce(0,:+)) /
    (Float(memo[:last_5].c\
ompact.length))
```

```
      memo[:sliding_averages].push memo_average
      memo
end[:sliding_averages]
```

Denormalize a Data Set

Sometimes, you have a normalized data set that you want to denormalize for human use.

For example, you may have a data set extracted from an SQL database. The first table is a list of names, uniquely identified by an ID, with each row formatted as {id,name}. (Note that there are two different Johns).

```
1,John
2,Jack
3,Jim
4,Jared
5,John
```

Then, you have a list of phone numbers with a unique ID, and a unique user ID, but phone numbers can be duplicated (because two people might share a phone); however, each entry is associated with a user. The format is {id, user_id, phone_number}. For example:

```
1,1,555-2344
2,1,555-1111
3,2,555-0000
4,3,555-0000
5,4,555-6666
7,5,555-7777
8,5,555-2344
```

However, you would like the denormalized list to look like this for easy user consumption:

```
John, 555-2344, 555,1111
Jack, 555-0000
Jim, 555-0000
Jared, 555-6666
John, 555-7777, 555-2344
```

This doesn't even follow the first normal form for data normalization, but that is OK for humans. First of all, you have already extracted your data sets into the arrays names and phones as follows:

```
irb> names
=> [[1,"John"],[2,"Jack"],[3,"Jim"],[4,"Jared"],[5,"John"]]
irb> phones
=> [[1,1,"555-2344"],[2,1,"555-1111"],[3,2,"555-0000"],
[4,3,"555-0000],[5,4,"555\
-6666"],[7,5,"555-7777"],[8,5,"555-2344"]]
```

Now, you want to match each phone number with each name:

```
cross_join = names.map do |name|
    phone.map do |phone|
        [name,phone]
    end
end
```

Then, you want to select only name/phone pairs where the user_id of the phone (the [1][1] index) matches the id of the name (the [0][0] index). For example:

```
name_phones = cross_join.select do |np|
    np[0][0] == np[1][1]
end
```

83

After that, you want to group by name, but you have to take the name_id because there are two Johns:

```
grouped_name_phones =
name_phones.reduce({}) do |namelist, np|
    if namelist[np[0]].nil?
        namelist[np[0]] = np[1][2]
    else
        namelist[np[0]].push np[1][2]
    end
    namelist
end
```

After that, you want to just save the name and the phone number in an array, as follows:

```
grouped_name_phones.map do |name,phones|
    [name[1],phones.map{|p|p[2]}.join(",")]
end
```

Pythagorean Triplets

The Pythagoras Theorem is an important one in mathematics. It states that, for a right triangle, the square of the hypotenuse is the sum of the squares of the other two sides. A Pythagorean triplet is one where all three sides of a right triangle are integers. For example, [3,4,5] or [5,12,13] or [8,15,17]. Math teachers the world over use these to create math tests so that they don't have to be bothered marking inexact long-division or square-root calculations.

How do you solve this problem of generating pythagorean triplets?

First of all, the hypotenuse is the longest side. So, we have to search all possible integer values of the hypotenuse. We'll start at 5, because we know that 5 is the length of the hypotenuse of the smallest Pythagorean triangle. We know that we have to return a triplet, which could be an array of three elements. We'll start with 5 and end with 17 for now:

```
(5..17).map do |hypotenuse|
  [nil,nil,hypotenuse]
end
=> [[nil, nil, 5], [nil, nil, 6], [nil, nil, 7], [nil, nil, 8],
[nil, nil, 9], [\
nil, nil, 10], [nil, nil, 11], [nil, nil, 12], [nil, nil, 13],
[nil, nil, 14], [\
nil, nil, 15], [nil, nil, 16], [nil, nil, 17]]
```

So far, so good. However, we need real numbers and not nil for the opposite and adjacent sides. We merely have to put them in nested loops. The opposite only has to search from 1 until the current hypotenuse, and the adjacent only has to search from 1 to the current opposite, as follows:

```
(5..17).map do |hypotenuse|
  (1..hypotenuse).map do |opposite|
    (1..opposite).map do |adjacent|
      [adjacent,opposite,hypotenuse]
    end
  end
end
=> [[[[1, 1, 5]], [[1, 2, 5], [2, 2, 5]], [[1, 3, 5],
[2, 3, 5], [3, 3, 5]], [[1\
, 4, 5], [2, 4, 5], [3, 4, 5], [4, 4, 5]], [[1, 5, 5],
[2, 5, 5], [3, 5, 5], [4,\
5, 5], [5, 5, 5]]], [[[1, 1, 6]], [[1, 2, 6], [2, 2, 6]],
[[1, 3, 6], [2, 3, 6]\
, [3, 3, 6]], [[1, 4, 6], [2, 4, 6], [3, 4, 6], [4, 4, 6]],
[[1, 5, 6], [2, 5, 6\
```

```
], [3, 5, 6], [4, 5, 6], [5, 5, 6]], [[1, 6, 6], [2, 6, 6],
[3, 6, 6], [4, 6, 6]\
, [5, 6, 6], [6, 6, 6]]], [[[1, 1, 7]], [[1, 2, 7], [2, 2, 7]],
[[1, 3, 7], [2, \
3, 7], [3, 3, 7]], [[1, 4, 7], [2, 4, 7], [3, 4, 7], [4, 4, 7]],
[[1, 5, 7], [2,\
5, 7], [3, 5, 7], [4, 5, 7], [5, 5, 7]], [[1, 6, 7], [2, 6, 7],
[3, 6, 7], [4, \
6, 7], [5, 6, 7], [6, 6, 7]], [[1, 7, 7], [2, 7, 7], [3, 7, 7],
[4, 7, 7], [5, 7\
, 7], [6, 7, 7], [7, 7, 7]]], [[[1, 1, 8]], [[1, 2, 8], [2, 2,
8]], [[1, 3, 8], \
[2, 3, 8], [3, 3, 8]], [[1, 4, 8], [2, 4, 8], [3, 4, 8], [4, 4,
8]], [[1, 5, 8],\
[2, 5, 8], [3, 5, 8], [4, 5, 8], [5, 5, 8]], [[1, 6, 8], [2, 6,
8], [3, 6, 8], \
[4, 6, 8], [5, 6, 8] , [6, 6, 8]], [[1, 7, 8], [2, 7, 8], [3,
7, 8], [4, 7, 8], \
[5, 7, 8], [6, 7, 8], [7, 7, 8]], [[1, 8, 8], [2, 8, 8], [3, 8,
8], [4, 8, 8], .\
. . . . . . . . . . . . . . . . .
```

The returned value is a mess, but it gets all the reasonable combinations. Now, once we have all of the numbers, we don't want to keep numbers that aren't triplets; we only keep the triplets. That means that we should only choose values for "opposite" that make a Pythagorean triplet with the other two. In other words, the last call to map should be preceded by a select where the predicate is the Pythagorean triplet rule:

```
(5..17).map do |hypotenuse|
  (1..hypotenuse).map do |opposite|
    (1..opposite).select do |adjacent|
```

```
    adjacent*adjacent + opposite*opposite ==
    hypotenuse*hypotenuse
  end.map do |selected_adjacent|
    [selected_adjacent,opposite,hypotenuse]
  end
 end
end
=> [[[], [], [], [[3, 4, 5]], []], [[], [], [], [], [], []],
[[], [], [], [], []\
, [], []], [[], [], [], [], [],    [], [], []], [[], [], [],
[], [], [], [], []\
, []], [[], [], [], [], [], [], [], [[6, 8, 10]], [], []], [[],
[],     [], [], \
[], [], [], [], [], [], []], [[], [], [], [], [], [], [], [],
[], [], [], []], [\
[], [], [], [], [], [],     [], [], [], [], [], [[5, 12, 13]],
[]], [[], [], []],\
[], [], [], [], [], [], [], [], [], [], []], [[], [],
[],    [], [], [], [], [\
], [], [], [], [[9, 12, 15]], [], [], []], [[], [], [], [], [],
[], [], [], [], \
[], [], [],      [],[], [], []], [[], [], [], [], [], [], [],
[], [], [], [], [\
], [], [], [[8, 15, 17]], [], []]]
```

This looks better, but we have a lot of blank arrays, and they are nested three deep. That is because map always returns an array. So, nested map calls will return arrays of arrays. The first thing to do is try to flatten the outer array, as follows:

```
(5..17).map do |hypotenuse|
  (1..hypotenuse).map do |opposite|
    (1..opposite).select do |adjacent|
```

```
        adjacent*adjacent + opposite*opposite ==
        hypotenuse*hypotenuse
    end.map do |selected_adjacent|
        [selected_adjacent,opposite,hypotenuse]
    end
  end
end.flatten
=> [3, 4, 5, 6, 8, 10, 5, 12, 13, 9, 12, 15, 8, 15, 17]
```

This is not right, because we went too far in flattening the array of arrays of arrays. If we group them in triplets, we actually have the Pythagorean triplets from 5 to 17 for the hypotenuse. However, we want actual triplets preserved. So, we need to successively flatten the results of each map. In Ruby, there is a function called flat_map that does this. We need to flatten the outer map and the first inner map in order to have an array of triplet arrays. I'll also bump up the number to 26:

```
(5..26).flat_map do |hypotenuse|
  (1..hypotenuse).flat_map do |opposite|
    (1..opposite).select do |adjacent|
        adjacent*adjacent + opposite*opposite ==
        hypotenuse*hypotenuse
    end.map do |selected_adjacent|
        [selected_adjacent,opposite,hypotenuse]
    end
  end
end
=> [[3, 4, 5], [6, 8, 10], [5, 12, 13], [9, 12, 15], [8, 15,
17], [12, 16, 20], \
[15, 20, 25], [7, 24, 25], [10, 24, 26]]
```

This is the result we were seeking. You can bump the number up higher to see what happens.

CHAPTER 4

Reverse-Engineering Complex Solutions

At some point, you will be presented with code that was written by others (or by you, a long time ago, and you have since forgotten how it works), and you will need to understand what the code did and if it needs to be fixed or evolved.

Reading a data-processing workflow that uses map, reduce, or select is very different from reading typical code written in an imperative coding style. In addition, Ruby's way of passing blocks can be confusing to many people unfamiliar with that coding style. I have experienced people reading my map/reduce/select data-processing cascades and reacting with anger, confusion, and scorn. And I myself have felt that way about such code from other people. I switched from imperative coding to using this style because I can solve my problems with fewer lines of code, and I can solve the problems in a disciplined way because each step in the cascade is individually debuggable.

In this section, I will present an example and walk you through reverse-engineering the solution and gaining understanding of the techniques used to derive the solution in the first place.

© Jay Godse 2018
J. Godse, *Ruby Data Processing*, https://doi.org/10.1007/978-1-4842-3474-7_4

Solution for Pascal's Triangle

I found the following code on the internet (at `http://kellegous.com/j/2006/06/28/post-9/`) for a Pascal's triangle (I picked 0..5 to make it easier to display). It is used here with the permission of the code's author, Kelly Norton.

```
text =
(0..5).reduce([[1]]) { |a,x|
    a.unshift a.first.reduce([0,[]]) { |b,y|
        [y,b.last << (b.first + y)]
    }.last + [1]
}.reduce([]) { |c,z|
    next [z[z.length/2].to_s.length*2,z.length,""] if c.empty?
    [c[0],c[1], z.map { |j|
            j.to_s.center(c[0])
        }.join('').center(c[0]*c[1])+"\n#{c.last}"]
}.last
puts text
```

When I ran it, I got the following:

```
            1
          1   1
        1   2   1
      1   3   3   1
    1   4   6   4   1
  1   5   10   10   5   1
=> nil
```

Lovely. It looks like a nicely formatted Pascal's triangle, and it is a lot more concise than the example I wrote.

This function seems to be using two cascading reduce steps to build the result. Let's see what the first stage does:

```
text =
(0..5).reduce([[1]]) { |a,x|
    a.unshift a.first.reduce([0,[]]) { |b,y|
        [y,b.last << (b.first + y)]
    }.last + [1]
}

=> [[1, 6, 15, 20, 15, 6, 1], [1, 5, 10, 10, 5, 1], [1, 4, 6,
4, 1], [1, 3, 3, 1\
], [1, 2, 1], [1, 1], [1]]
```

This is interesting. It looks like the Pascal's triangle array, but in reverse. Let's add a debugging statement. Don't forget to return the memo afterward.

```
text =
(0..5).reduce([[1]]) { |a,x|
    memo =
    a.unshift a.first.reduce([0,[]]) { |b,y|
        [y,b.last << (b.first + y)]
    }.last + [1]
    puts "After processing x: #{x}, the memo is: #{memo}"
    memo
}

After processing x: 0, the memo is: [[1, 1], [1]]
After processing x: 1, the memo is: [[1, 2, 1], [1, 1], [1]]
After processing x: 2, the memo is: [[1, 3, 3, 1], [1, 2, 1],
[1, 1], [1]]
After processing x: 3, the memo is: [[1, 4, 6, 4, 1],
[1, 3, 3, 1], [1, 2, 1], [\
```

1, 1], [1]]
After processing x: 4, the memo is: [[1, 5, 10, 10, 5, 1],
[1, 4, 6, 4, 1], [1, \
3, 3, 1], [1, 2, 1], [1, 1], [1]]
After processing x: 5, the memo is: [[1, 6, 15, 20, 15, 6, 1],
[1, 5, 10, 10, 5,\
1], [1, 4, 6, 4, 1], [1, 3, 3, 1], [1, 2, 1], [1, 1], [1]]
=> [[1, 6, 15, 20, 15, 6, 1], [1, 5, 10, 10, 5, 1],
[1, 4, 6, 4, 1], [1, 3, 3, 1\
], [1, 2, 1], [1, 1], [1]]

We can see that each run of reduce adds a row to the triangle and saves the triangle in the memo. Let's see what happens in each of the inner reduce steps by printing out key data after each main step:

```
text =
(0..5).reduce([[1]]) { |a,x|
    puts "Before processing a: #{a}   , x: #{x}"
    memo =
    a.unshift a.first.reduce([0,[]]) { |b,y|
        puts " Before processing inner loop, b: #{b}, y: #{y}"
        inner_memo =
        [y,b.last << (b.first + y)]
        puts "  After processing inner loop y: #{y}, the inner
        memo is #{inner_mem\
o}"
        inner_memo
    }.last + [1]
    puts "After processing x: #{x}, the memo is: #{memo}"
    puts ""
    memo
}
```

This yields a lot of output:

```
Before processing a: [[1]], x: 0
  Before processing inner loop, b: [0, []], y: 1
  After processing inner loop y: 1, the inner memo is [1, [1]]
After processing x: 0, the memo is: [[1, 1], [1]]

Before processing a: [[1, 1], [1]], x: 1
  Before processing inner loop, b: [0, []], y: 1
  After processing inner loop y: 1, the inner memo is [1, [1]]
  Before processing inner loop, b: [1, [1]], y: 1
  After processing inner loop y: 1, the inner memo is
  [1, [1, 2]]
After processing x: 1, the memo is: [[1, 2, 1], [1, 1], [1]]

Before processing a: [[1, 2, 1], [1, 1], [1]], x: 2
  Before processing inner loop, b: [0, []], y: 1
  After processing inner loop y: 1, the inner memo is [1, [1]]
  Before processing inner loop, b: [1, [1]], y: 2
  After processing inner loop y: 2, the inner memo is
  [2, [1, 3]]
  Before processing inner loop, b: [2, [1, 3]], y: 1
  After processing inner loop y: 1, the inner memo is
  [1, [1, 3, 3]]
After processing x: 2, the memo is: [[1, 3, 3, 1], [1, 2, 1],
[1, 1], [1]]

Before processing a: [[1, 3, 3, 1], [1, 2, 1], [1, 1], [1]],
x: 3
  Before processing inner loop, b: [0, []], y: 1
  After processing inner loop y: 1, the inner memo is [1, [1]]
  Before processing inner loop, b: [1, [1]], y: 3
  After processing inner loop y: 3, the inner memo is
  [3, [1, 4]]
```

Before processing inner loop, b: [3, [1, 4]], y: 3
After processing inner loop y: 3, the inner memo is
[3, [1, 4, 6]]
Before processing inner loop, b: [3, [1, 4, 6]], y: 1
After processing inner loop y: 1, the inner memo is
[1, [1, 4, 6, 4]]
After processing x: 3, the memo is: [[1, 4, 6, 4, 1],
[1, 3, 3, 1], [1, 2, 1], [\
1, 1], [1]]

Before processing a: [[1, 4, 6, 4, 1], [1, 3, 3, 1], [1, 2, 1],
[1, 1], [1]], x\
: 4
Before processing inner loop, b: [0, []], y: 1
After processing inner loop y: 1, the inner memo is [1, [1]]
Before processing inner loop, b: [1, [1]], y: 4
After processing inner loop y: 4, the inner memo is
[4, [1, 5]]
Before processing inner loop, b: [4, [1, 5]], y: 6
After processing inner loop y: 6, the inner memo is
[6, [1, 5, 10]]
Before processing inner loop, b: [6, [1, 5, 10]], y: 4
After processing inner loop y: 4, the inner memo is
[4, [1, 5, 10, 10]]
Before processing inner loop, b: [4, [1, 5, 10, 10]], y: 1
After processing inner loop y: 1, the inner memo is
[1, [1, 5, 10, 10, 5]]
After processing x: 4, the memo is: [[1, 5, 10, 10, 5, 1],
[1, 4, 6, 4, 1], [1, \
3, 3, 1], [1, 2, 1], [1, 1], [1]]

Before processing a: [[1, 5, 10, 10, 5, 1], [1, 4, 6, 4, 1],
[1, 3, 3, 1], [1, 2\

, 1], [1, 1], [1]], x: 5
  Before processing inner loop, b: [0, []], y: 1
  After processing inner loop y: 1, the inner memo is [1, [1]]
  Before processing inner loop, b: [1, [1]], y: 5
  After processing inner loop y: 5, the inner memo is
  [5, [1, 6]]
  Before processing inner loop, b: [5, [1, 6]], y: 10
  After processing inner loop y: 10, the inner memo is
  [10, [1, 6, 15]]
  Before processing inner loop, b: [10, [1, 6, 15]], y: 10
  After processing inner loop y: 10, the inner memo is [10,
  [1, 6, 15, 20]]
  Before processing inner loop, b: [10, [1, 6, 15, 20]], y: 5
  After processing inner loop y: 5, the inner memo is [5,
  [1, 6, 15, 20, 15]]
  Before processing inner loop, b: [5, [1, 6, 15, 20, 15]], y: 1
  After processing inner loop y: 1, the inner memo is [1,
  [1, 6, 15, 20, 15, 6]]
After processing x: 5, the memo is: [[1, 6, 15, 20, 15, 6, 1],
[1, 5, 10, 10, 5,\
1], [1, 4, 6, 4, 1], [1, 3, 3, 1], [1, 2, 1], [1, 1], [1]]

=> true

What is apparent is that each step of the inner loop iterates over the
last row (the first element of a[]) and unshifts it onto the inner memo
array (by sliding everything over to the right). It then adds that number
(y) to the first element of b[] (which is always an integer or nil) and then
pushes it onto the end of the last element of b[] (which is always an array).
The resulting inner memo after processing is a number as the first element
and a partial row as the second element. The first element is needed for
the next iteration when it is added to the last element of the partial row.

After the inner loop runs for the last time, the second element of the memo is [1,6,15,20,16,5], which is almost a full row of the triangle. That is why the result of the inner loop has [1] pushed onto the end of the array, which makes a full row of the triangle.

From here, one only has to reverse the result to get the raw data for the triangle and then print it out, as I did earlier in this book with the example I wrote. I leave it as an exercise for the reader to see how this example formats and prints the triangle in a nice symmetrical format.

Throughout this book, I have tried my best to walk you through the mechanics of using map, reduce, and select to solve data processing problems. If you have gone through this book, and typed in the examples and pondered the results, I sincerely hope that it has stimulated your thinking about how to apply these to your own data processing problems. I also hope that you can carry these techniques forward to solve your own data processing problems in the future.

Index

A, B, C, D, E

Data sets
  area code, 76
  denormalize, 82–84
  list of names, 42
  lookup policy data, 44–45
  sliding averages, 77–81
  uniq(), 43
Debugging
  map, 25–26
  reduce, 26
  select, 27

F, G, H

FizzBuzz, 27, 29–30

I, J, K, L

Intersection points
  clusters, 71, 73–75
  functions, 67–68, 70
  map(), 70–71

M, N, O

Map
  description, 13–14
  list of names, 33, 35

name bar, 63–66
names to CSV, 36–38
odd cubes, 30–32
random list of
  names, 39–41
sequence of sales, 46–47

P, Q

Pascal's triangle, 48–55, 90–95
Postfix/reverse polish notation
  command line, 60–62
  stack, 56–58
Pythagoras Theorem, 84–88

R

Reduce
  list of names, 33, 35
  max, 21–22
  memo/element, 16–18
  name bar, 63–66
  names to CSV, 36–38
  odd cubes, 30–32
  random list of
    names, 39–41
  reverse, 20–21
  sequence of sales, 46–47
  uniq, 19–20

S, T, U, V, W, X, Y, Z

Get the eBook for only $5!

Why limit yourself?

With most of our titles available in both PDF and ePUB format, you can access your content wherever and however you wish—on your PC, phone, tablet, or reader.

Since you've purchased this print book, we are happy to offer you the eBook for just $5.

To learn more, go to http://www.apress.com/companion or contact support@apress.com.

Apress®

Printed in the United States
By Bookmasters